Agrarian Capitalism
and the World Market

Agrarian Capitalism and the World Market

Buenos Aires
in the Pastoral Age,
1840–1890

Hilda Sabato

University of New Mexico Press
Albuquerque

Library of Congress Cataloging-in-Publication Data

Sabato, Hilda.
Agrarian capitalism and the world market:
Buenos Aires in the pastoral age, 1840–1890
Hilda Sabato.
p.   cm.
Includes bibliographical references and index.
ISBN 0-8263-1218-7
1. Sheep industry—Argentina—Buenos Aires.
2. Wool industry—Argentina—Buenos Aires.
I. Title.
HD9436.A73B847   1990
338.1'76314—dc20
90-12635

A mi abuelo, José Añez

Contents

Illustrations

Figures

Maps

Tables

Acknowledgments

This book is an amended version of a doctoral thesis originally submitted to the University of London. Most of the research work was carried out at the Institute of Latin American Studies in London (1976–78) and at the Centro de Investigaciones Sociales sobre el Estado y la Administración (CISEA) in Buenos Aires (1978–80), with a grant from the Ford Foundation and additional financial assistance from CISEA, the British Council, and the University of London Central Research Fund. After five years of pursuing other research interests, I returned to my old intellectual obsessions and set to work on the elaboration of the present text, which was written at CISEA during 1986. Financial support for this stage came from the Argentine Consejo Nacional de Investigaciones Científicas y Técnicas. I am indebted to those institutions as well as to my friends, colleagues, and all those others who in one way or another helped me in this endeavor.

My particular gratitude goes to Professor John Lynch, who supervised all stages of my research work, for his help and advice, and to Dr. Tulio Halperin Donghi, who encouraged me to write this book, for his intellectual counsel and friendly support. I also express my gratitude to Juan Carlos Korol, with whom I spent long hours discussing my ideas, as well as to Leandro Gutiérrez, Luis Alberto Romero, Beatriz Sarlo, and my other colleagues at PEHESA (Programa de Estudios de Historia Económica y Social Americana) and CISEA for their unfailing support.

So many other people helped me at various stages of my research that it would be impossible to name them all here. However, I would particularly like to thank Luis Ortega for his encouragement throughout; Eric Hobsbawm, Horacio Giberti, and Jorge F. Sábato for their suggestions and comments in the formative stages of my work; Ernesto Laclau not only for discussing with me some of the initial ideas around the subject of this book but also for saving me months of archival work by handing to me material he had gathered in primary sources; the members of the jury of the José Luis Romero International History Prize—namely Drs. Tulio Halperin Donghi, Richard Morse, Juan A. Oddone, and Gregorio Weinberg—for granting me second prize in the competition held in 1981 and for encouraging me to turn my thesis into a book; Gilbert Merkx and Donna Guy for helping me to find a publisher; Messrs. George Cormouls–Houlès, Claude Fohlen, Hilda Herzer, Charles Jones, Alfredo Montoya, Emilio Roux, Karel Veraghtert, and Stuart Wright for the information they kindly provided me; the librarians and other personnel at the Archivo del Banco de la Provincia de Buenos Aires, Archivo General de la Nación, Biblioteca de la Sociedad Rural Argentina, Biblioteca Tornquist, British Library, CEDES, CISEA, Ets. Cormouls–Houlès—Père et fils, Federación Lanera Argentina, and Institute of Latin American Studies for their helpful assistance, and Inés Queirolo and Cristina Durmüller for typing this long and demanding manuscript.

I also wish to express my warmest acknowledgments to all members of my family, and most specially to my husband Carlos Reboratti for bearing my changing moods with incredible endurance, and to my sons Andrés and Julián for fortitude beyond their tender years. Without my family's help and encouragement this book would truly never have been written.

Finally, I apologize to my readers for having presumed to write this text in English, a language which obviously is not my mother tongue.

Introduction

The oil springs of Pennsylvania are outdone and the gold
fields of California and Australia eclipsed, by the sheep-
farming business of Buenos Aires.
The Brazil and River Plate Mail, 22 November 1865.

In 1840 few Argentines could have imagined the changes their
country would undergo in the following four decades. In fact,
most of them did not even think of themselves as Argentines.
Regional economies, provincial societies, and political atomiza-
tion—such was their world in the first half of the nineteenth
century. By 1880, however, a new reality had been forged. In just
these few decades, the country that was to constitute the Argentine Republic had undergone an accelerated process of change.
In forty years it had multiplied its population fivefold, and the
volume of its trade with the rest of the world had increased seven-
fold. After four decades of internal struggle and external wars, it
had become politically organized as a nation, and had defined
the basic features of its social and economic structures for a long
time to come. At the moment of reshuffling within the international market, it consolidated its place as supplier of foodstuffs
and raw materials, and as a consumer of manufactures, capital,
and men.

The province of Buenos Aires was at the very heart of this
transformation. Sparsely populated and only partially exploited
as it was before mid-century, only a few decades later it was to
become a flourishing area of agricultural production. Always
struggling to impose its predominance upon the rest of the provinces but only partially succeeding in the first half of the nineteenth century, by 1880 Buenos Aires had risen to undisputed
leadership. Not only had it played a key role in the definition of
the sociopolitical order, crowned by that symbolic year of the

1

Organización Nacional, but also it had drawn the basic lines for the pattern of development the country would follow in the future. Within the province, this period was marked by unprecedented growth, built upon the expansion and consolidation of capitalism. Although this process had its starting point earlier in the history of the River Plate, it had acquired new impulse and vigor after 1850, with the growth of wool production and foreign trade.

Under the stimulus of an increasing international demand for wool, sheep raising became the most lucrative business and the main source of wealth, while it gave Buenos Aires the opportunity of participating fully in an international market governed by the rules of free trade and comparative advantages. Therefore, it is not surprising to find that sheep raising expanded rapidly in the province, pushing cattle southward, invading the best land, attracting capital and men, and arousing new hopes and ambitions.

Very soon, the pastoral industry became the leading sector in the process of accumulation. How this was achieved is the main question this work attempts to answer. The purpose of this study is to analyze the development of the sheep-raising industry between the late 1840s and the 1880s, concentrating on how sheep raising and wool production and trade were organized in Buenos Aires.

Although the period under study has received the attention of historians and social scientists who have attempted general interpretations of the economic history of Argentina, or of particular aspects of it, very few works have concentrated on this period, and none of them have dealt specifically with the pastoral industry.[1]

From different standpoints, and within slightly different chronological limits, most authors agree that these decades were decisive in terms of the full incorporation of the country into the world market, and that sheep raising was the dynamic sector which speeded up that process.

Few interpretations see the decades between the downfall of Rosas and the advent of Roca to the presidency either as the breaking point between two fundamentally different orders or as only one stage in a continuous process of incorporation in the world market and internal adjustment to that process. However, if the old view of Ingenieros, who saw the period as one of transition from feudalism to a more modern economic system (*agro-pecuario* for Ingenieros, *capitalist* for more recent historians) has been superseded, the idea that the systems were fundamentally different has not. Thus, scholars like Ortiz and Giberti tend to stress that difference, even referring to a deep antagonism within the propertied classes between cattle and sheep *estancieros*. On the other hand, continuity is emphasized in a study by Brown, who considers wool as just one more in the list of staples whose production since colonial times has defined the country's pattern of growth.[2]

Avoiding these opposing interpretations, most recent works—those of Ferrer, Gorostegui and Halperin among others—are careful to point out both continuities and ruptures between the pre- and post-Rosas orders, but agree in stressing the dynamism of the process of growth inaugurated in the pastoral era.[3]

There are no such differences in the interpretation of the transition from that era to the period of unparalleled expansion that followed. Most scholars see the former as the first step in the achievement of the latter, and point to the continuities between both patterns of accumulation. Differences arise, however, when these patterns are analyzed. Thus, although most works agree as to the role played by the country's incorporation in the world market in the process of growth, interpretations differ regarding both the consequences this incorporation had for the country's development and future performance, and the internal factors that conditioned the whole process.

Two main interpretations should be mentioned, which I will refer to as the *optimistic* and the *pessimistic* versions. The optimistic interpretation stresses the spectacular process of growth experienced by Argentina during the decades of the export-led economy, and tries to demonstrate that the country made optimal use of its resources and comparative advantages, the result being the development of a capitalist economy ruled by the laws of the market which stimulated an optimal use of factors. Thus the problems accruing after 1930 cannot be traced to the previous decades and have to be explained by analyzing economic policies and performance since the Great Crisis.[4]

The pessimistic interpretation explores the limitations on the process of growth inaugurated in the last decades of the nine-

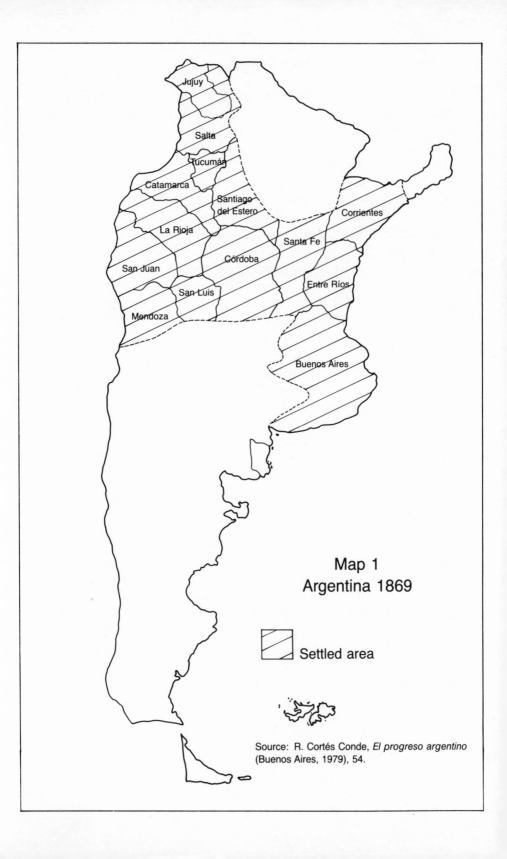

Map 1
Argentina 1869

Settled area

Source: R. Cortés Conde, *El progreso argentino* (Buenos Aires, 1979), 54.

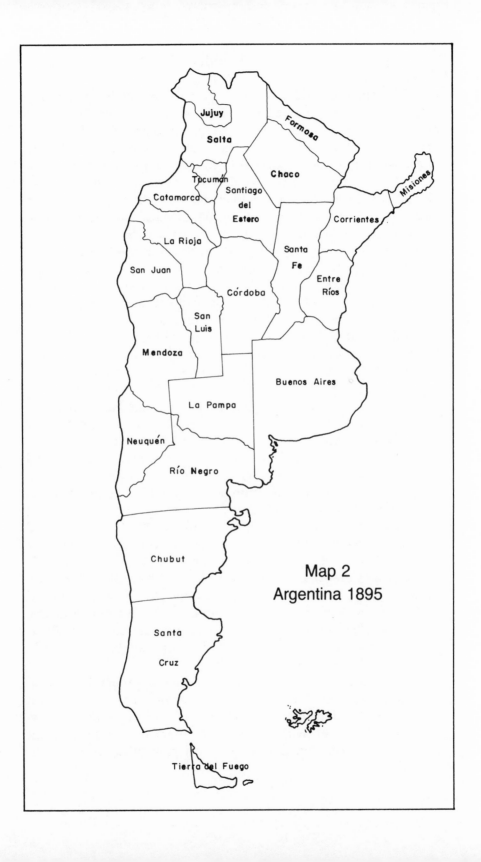

Map 2
Argentina 1895

teenth century particularly bearing on the possibilities of developing an autonomous and truly capitalistic economy. A number of works adopt this general perspective, and although they differ greatly in dealing with the specific issues involved, most of them stress the influence of external factors in conditioning the country's development, as the local economy became strongly dependent on the international market and the metropolitan countries. They also pay attention to the internal obstacles to development, particularly underlining the problems posed by a land tenure system which is seen as a negative factor in terms of the consolidation of a truly capitalist system.[5]

In the late 1960s and early 1970s these topics became key issues in a more general debate that swept academic and political circles linked to the tradition of the left in Latin America—this was the dependency debate. A wide variety of works were inspired by the interpretation initially put forward by André Gunder Frank and which found a more subtle formulation in the hands of Fernando H. Cardoso and Enzo Faletto.[6] Reacting against the different theories that in one way or another postulated a parallel path of development for all nations in the world—thus anticipating for underdeveloped countries a delayed but certain road to industrialization and growth—the dependency approach questioned the basic assumptions upon which these expectations were built. In its simpler versions, this approach resulted in a series of works that predicted the impossibility of achieving development in a capitalist context, because integration into the world economy had meant for the countries in the periphery the consolidation of their role as satellites of the metropolises and a continuous process of surplus transference from those countries to the center. Moreover, in an attempt to explain current realities, these studies looked back into the history of Latin America and advanced a daring proposition that the countries of the region had been capitalistic since the moment in which they were incorporated into the world economy—that is, since colonial times.

These crude versions of the dependency approach have been subjected to criticism from different quarters, and by now most of their propositions have wisely been put aside by scholars. This approach, however, was also the framework for a number of sophisticated studies on the present and past of Latin American countries, studies whose findings shed new light on aspects of the problem that could not be "seen" before. This suggests that although dependency failed as a theory or as an overall explaining principle, it became a useful methodology for the study of specific cases, as has been maintained by Gabriel Palma.[7]

In the case of Argentine history, the dependency approach converged with the pessimistic versions which underlined the negative aspects of the process of growth inaugurated by the middle of the nineteenth century. Yet its initial versions were soon superseded by a new perspective put forward by Ernesto Laclau and Jorge F. Sábato.[8] Advancing beyond the arguments set by the different versions within the pessimistic tradition, they developed an interpretation that accounted for both the extraordinary growth experienced during the era of greatest expansion and the stagnation that followed. They found that the same factors which favored growth had later become obstacles to further expansion. Thus, in an article that included a devastating critique of Gunder Frank's dependency arguments, Laclau finds that the main feature of Argentina's "dependent capitalism" was a pattern of growth and distribution of wealth that relied heavily on the transference of international differential rent.

In much the same way as today's oil-exporting countries, at the turn of the century Argentina found in its primary production a source of rent. In this case, the fertility of the soil was the cause of significant differences between prices of production and market prices of the staples exported, which in turn led to a net flow of resources from consuming to producing areas. In Laclau's words: "In our economy, the expansion of rent took the place that in non-dependent capitalism corresponds to capitalist accumulation."[9] He further argues that limitations to the process of capitalist development came precisely from this peculiar pattern of growth, in which the landowning oligarchy was to play the dominant role. By maintaining that rent was the main source of wealth for the country, Laclau introduced a new and controversial argument into the ongoing debate on the history of Argentina's economic and social development. Yet, by considering the landowning class to be the main recipient of that wealth, he came to terms with the most traditional views on the subject.

The work of Jorge F. Sábato presents a new set of arguments

to challenge these views. Without questioning the hypothesis on the role of international Ricardian rent for the country, he points out that the existence of such rent does not necessarily imply that landowners were the main beneficiaries of the export economy or the dominant group in society. In fact, he advances a different interpretation of the ruling classes. In his view, more important than the ownership of land was the control of commerce and finance that gave these classes the basic instruments for concentrating wealth and power. Their drive toward land and rural production should be seen as a consequence of the dynamics of their original, and central, activities in the commercial and financial world, which provided the framework for their economic behavior. Obviously, this attractive proposition challenges the traditional interpretations that stress the role of the landowning class in defining the country's pattern of development, and it sheds new light on the history of Argentina's expansive phase, as well as on the uneven course followed by its economy since the 1930s.

Although many of the questions raised by this interesting controversy on the nature of Argentine society are at the background of this present study, far from attempting to resolve them or to forward a general interpretation of the development of capitalism, this work inquires into certain specific aspects of that process in the hope of contributing to the debate. As a point of departure, this study acknowledges Laclau's hypothesis on the existence of an international differential rent in favor of Argentina as a consequence of its comparative advantages in producing primary goods due to the extraordinary fertility of the soil in the pampean region. Wool was probably the first of the staples to be exported under such favorable conditions.[10] Therefore, this work refers to a key stage in the incorporation of Argentina into the world market. It will not, however, explore the question of rent, but will turn to the problem of how production and trade were organized in this expanding export economy.

The production of staples in Argentina attracted scholars in the past, but most works focus upon the age of mixed farming, when grain and cattle became the main exports of the country.[11] This emphasis has illuminated the period of greatest expansion of the Argentine economy, inaugurated at the turn of the century, when the pampean region was already well set in the way of rural production and well established in the international market as a supplier of primary goods and as a consumer of capital and manufactures. The decades that preceded this extraordinary expansion paved the way to such an achievement, and it was then that the basic features of the future economic and social structure were drawn. This study argues that the pastoral era was a crucial stage in the consolidation of a capitalist economy in the region, because it was during the second half of the nineteenth century that the process of capitalist accumulation gained momentum, under the lead of the pastoral sector. This process involved the creation of a land and a labor market; the organization of production on the basis of wage labor; the expansion of a new type of rural enterprise, the sheep estancia; and the consolidation of a class of capitalist-landlords, who combined landownership and capitalist entrepreneurship in the organization of their rural concerns.

Yet the direction followed by this process was greatly influenced by the fact that the main stimulus for growth lay in the international scene. This work will pay attention both to the direct impact of this close relationship with the world market upon the organization of rural production, and to the effect of international capital in the development of the export economy.

Actually, during the pastoral age, foreign investments were mainly channeled into government loans and the construction of railways. However, they also played a key role in the expansion of the commercial and financial networks that were to provide for the increasing requirements of the wool trade. Traditionally, it has been held that during the long decades in which Argentina was closely linked to the international market, these networks were monopolized by foreign capital, while production was left in the hands of the local classes.[12] Without denying the importance of European interests in the control of finance and trade, we will stress the participation of native capital in most sectors of the export economy. In this sense, the period bears witness to a tight intertwining of rural and urban interests, and it is quite clear that in the top ranks of the propertied classes success was associated with the ability to combine a strong involvement with the commercial and financial world with an increasing concern for rural production. This fact suggests that the formation of the ruling classes was already following the path so perceptively unveiled by Jorge Sábato.

Within this general framework, and leaving aside overall questions regarding the performance of the national economy and the structure of society, this study will explore the more limited territory of the pastoral industry in the River Plate. It will concentrate on the period between mid-century and the 1880s, when that industry became the leading and most dynamic sector in the Argentine economy, and it will specifically refer to the province of Buenos Aires, as this area was by far the main producer and exporter of wool throughout the second half of the nineteenth century. During this period of consolidation of a national market and a centralized state, the province of Buenos Aires found different ways of insuring its relative autonomy from the rest of the country, with the province even resorting to political secession when that autonomy was in peril. Therefore, to a certain extent it is possible to speak of a local capitalist development, and although most of the problems discussed in this work are placed in the context of the country's economic and social history, the analysis of the agrarian structure and of wool production and trade is restricted to the area where the pastoral sector set the pace of capitalist accumulation.

This area was not the same, however, throughout the period under consideration. In the first place, the territory of the province expanded greatly as the Indians were pushed south and westward. Thus, in spite of temporary setbacks, the area included within the boundaries of the province increased from 139,624 km² in 1854, to 236,628 km² in 1879, and to 310,307 km² in 1881 (see Maps 3–5).[13] Secondly, the area under sheep raising also changed, and if by mid-century the counties immediately to the south of the city of Buenos Aires were the first to turn pastoral, this pattern later expanded into the west and north, finally covering almost all the area north of the Salado River and not a few of the counties to the south of that river as well.

Although most of this work refers to sheep raising and the wool trade in the whole province, it concentrates on those counties which shared certain basic ecological features and at the same time were persistently pastoral. These are the counties north of the Salado River, excluding only those of the agricultural belt of

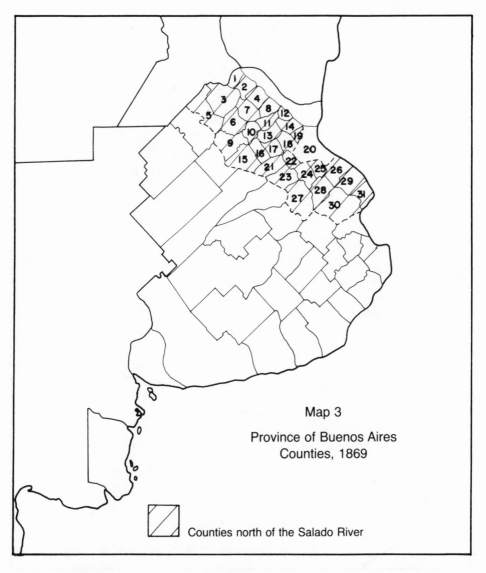

Map 3

Province of Buenos Aires
Counties, 1869

Counties north of the Salado River

1	San Nicolás	12	Zárate	22	Las Heras
2	Ramallo	13	San Andrés de Giles	23	Lobos
3	Pergamino	14	Exaltación de la Cruz	24	Cañuelas
4	San Pedro	15	Chivilcoy	25	San Vicente
5	Rojas	16	Suipacha	26	Ensenada
6	Salto	17	Mercedes	27	Monte
7	Arrecifes	18	Luján	28	Ranchos
8	Baradero	19	Pilar	29	Magdalena
9	Chacabuco	20	Buenos Aires and	30	Chascomús
10	Carmen de Areco		surrounding counties	31	Rivadavia
11	San Antonio de Areco	21	Navarro		

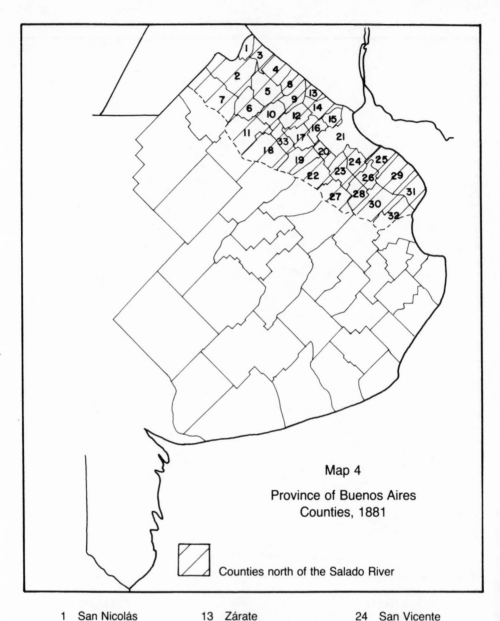

Map 4

Province of Buenos Aires
Counties, 1881

Counties north of the Salado River

1	San Nicolás	13	Zárate
2	Ramallo	14	Exaltación de la Cruz
3	Pergamino	15	Pilar
4	San Pedro	16	Luján
5	Rojas	17	Mercedes
6	Salto	18	Chivilcoy
7	Arrecifes	19	Navarro
8	Baradero	20	Las Heras
9	Chacabuco	21	Buenos Aires and
10	Carmen de Areco		surrounding counties
11	Chacabuco	22	Lobos
12	San Andrés de Giles	23	Cañuelas

24	San Vicente
25	Ensenada
26	Brandzen
27	Monte
28	Ranchos
29	Magdalena
30	Chascomús
31	Rivadavia
32	Viedma
33	Suipacha

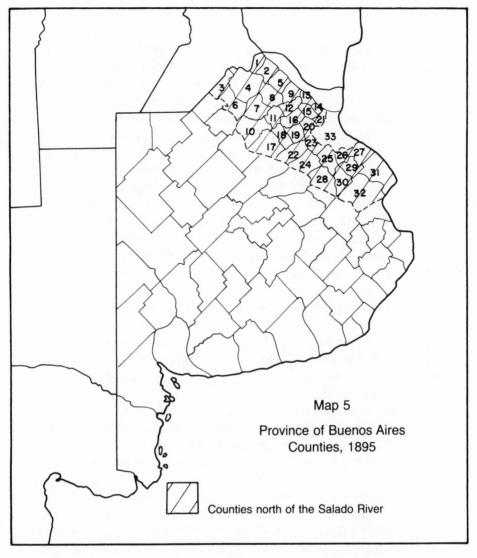

Map 5

Province of Buenos Aires
Counties, 1895

Counties north of the Salado River

1	San Nicolás	12	San Antonio de Areco	23	Las Heras
2	Ramallo	13	Zárate	24	Lobos
3	Colón	14	Campana	25	Cañuelas
4	Pergamino	15	Exaltación de la Cruz	26	San Vicente
5	San Pedro	16	San Andrés de Giles	27	Ensenada
6	Rojas	17	Chivilcoy	28	Monte
7	Salto	18	Suipacha	29	Brandzen
8	Arrecifes	19	Mercedes	30	Ranchos
9	Baradero	20	Luján	31	Magdalena
10	Chacabuco	21	Pilar	32	Chascomús
11	Carmen de Areco	22	Navarro	33	Buenos Aires and surrounding counties

Buenos Aires. Besides occupying the best lands, most of these counties had been situated within the frontier since the 1820s, and with more or less success they had been settled by cattle estancieros before the pastoral age.[14] Therefore, the area presents a relatively homogeneous geographical and historical point of departure for analyzing the transformations experienced by the agrarian structure in the period under study. Moreover, this was a key region in the definition of the rural transformations that were a recurrent and decisive feature of the Argentine economy. From the initial stages of the old cattle estancia, through pastoral times to the later phase of mixed farming, this area was always the first to witness the symptoms of change, leading the process of transformation that later was to extend to the rest of the Pampas. This work, then, refers throughout to the pastoral sector as a whole as well as to its development north of the Salado River.

The expansion of sheep raising is described in Chapter 1, which also analyzes the fluctuations of the pastoral industry throughout the period. Land was one of the keystones of that expansion and determined the way in which it was achieved, and Chapter 2 deals with land distribution and the structure of property in Buenos Aires during the pastoral era.

In Argentina, land has been a controversial issue ever since the future of the large extension of public lands was called into question. Publicists and politicians in the nineteenth century understood that issue to be closely related to the problem of the kind of society they envisaged for their country. Initially, most of the land was public land. It had belonged first to the Spanish Crown and later to the state, thus coming under the control of the federal government or the provincial administrations. Different solutions were put forward in relation to these lands. In terms of government policies, laws inspired by the American pattern of a farmer society, which meant stating limits in terms of concentration, alternated with "pragmatic" measures which favored privatization without limitations, thus benefiting the most powerful. At the end of the road, attempts to found a local "Midwest" had failed, and by the last decades of the century, land concentration was the result of the long process of transference of public lands into private hands. The extraordinary economic expansion achieved in the following decades contributed to appease controversy around

the land issue, but critical opinions always managed to survive—
even if isolated and marginal.[15]

These two basic standpoints—one critical, the other pragmatic—still persist when dealing with the history of Argentina's
land tenure system. The heirs of the pragmatic perspective claim
that concentration was inevitable; that extensive use of one factor—land—was necessary where labor and capital were scarce, in
order to ensure profitability. This interpretation underlines the
role of the market in the distribution of land, minimizing the
importance of institutional and political factors in the definition
of the land tenure pattern. Capitalist rationality on the part of
rural entrepreneurs and the free operation of supply and demand
forces in the market are basic assumptions in this argument.[16]

The critical perspective has an altogether different interpretation of the origins of the land tenure system: stressing the importance of the initial distribution of public lands which favored
a privileged few. Control over such a key resource turned these
sectors into a landowning oligarchy. This original process had a
decisive impact on the development of Argentine society. By concentrating so much power in the hands of a class that based its
fortune upon the control of land and the accumulation of rent, it
hindered proper capitalist development.[17]

In dealing with the problem of land, I have chosen to leave
aside both perspectives in order to analyze the process of effective
appropriation and incorporation into productive use of the territory of the province during the second half of the nineteenth
century. Consolidation of private property, transference of public
land into private hands, and the expansion of the frontier were
the means to that end, with the state playing a key role in this
process, particularly in the initial distribution of public lands which
favored a relatively small and privileged social group. In Buenos
Aires, this drive toward land was parallel to the expansion of
production, so that those who benefited from land grants and
cheap sales by the state enjoyed the advantages of an initial monopoly of that resource. Yet, simultaneously, the market widened
its field of operations because, once in private hands, land could
be sold or purchased, and supply and demand depended more
and more on the private decisions of individuals.

What was the result of this peculiar combination of public supply and market operation in terms of the structure of property in this period? Chapter 2 will seek an answer to this question, by analyzing the way this pattern changed between 1836 and 1890, and by exploring why, although subdivision was an important feature during this period, the structure of property did not experience substantial changes, as extensive holdings and enterprises remained predominant.

A second keystone in the process of capitalist organization was the consolidation of a free labor market, and Chapter 3 analyzes this process. Labor has received less attention than land, although in this case it is also possible to mention a pragmatic and a critical perspective. And while the former concentrates on the operation of the market, stressing how supply and demand were equilibrated, the latter emphasizes the process of formation of that market. Thus, it concentrates on the coercive mechanisms enforced to create a free labor supply and studies the consequences of a very unstable demand upon the labor force that was effectively incorporated into the market.[18]

In this work I have endorsed the critical perspective of labor. After analyzing the changes in the patterns of labor demand observed since mid-century, Chapter 3 deals with the creation of a free labor supply meant to meet the demands posed by the expanding economy. There were two sources for this supply: immigrants, who arrived in large numbers during these decades of expansion and were soon to become the main source of labor supply; and native workers, who had enjoyed alternative ways of subsistence other than wage labor, and at this stage were driven into the market.

Within the pastoral sector, the relationship between capital and labor adopted three main forms—wage labor, sharecropping, and family labor. In Chapter 3, these forms are approached from the point of view of the development of a free labor market, while in Chapters 4 and 5 they are viewed in their combination within estancias and farms.

Chapters 4 and 5 refer to the prevailing types of productive units—the estancia and the sheep farm. Argentine historiography has devoted little attention to the organization of production, and although estancias are often mentioned, very few works deal with them as rural enterprises. The discussion of this point has been subordinated to the more general debate on the capitalist nature of Argentine rural economy—a debate that has revolved more around theoretical discussions than historical research and has paid too little attention to the organization of rural enterprises. The same thing happened with the estanciero. Interpretations abound: He is portrayed as a feudal landlord, a capitalist tycoon, and everything in between, yet very few works deal with his behavior as an entrepreneur. Finally, as regards farms, debate is almost nonexistent for the period and area under study, as it is assumed that this type of enterprise could never have existed in a rural environment dominated by the large enterprise, the estancia.

This book studies estancias and farms as units of production, analyzing their internal organization, performance, and perspectives as rural enterprises. The *estancia* is defined as a capitalist endeavor that produced pastoral goods for the market and whose main purpose was the maximization of benefits. For each individual enterprise, these benefits included not only profits but also rent, as most estancieros were also the owners of the holdings upon which the estancias were organized. Ownership of land and pastoral enterprise were one single business; through wage labor, sharecropping, and tenancy, estancieros received both surplus and rent, and the increase in the value of land was an important item in the process of accumulation. Chapter 4 studies the organization and economics of sheep estancias, as well as the main directions followed by the process of investment both within the enterprises and for the industry as a whole.

Sheep raising was carried out not only in these estancias, but also in *sheep farms*, which were family concerns whose main purpose was to produce pastoral goods to sell in the market at a profit, relying mainly on family labor and only occasionally making use of wage labor. Chapter 5 deals with the emergence of this type of enterprise during the decades of pastoral expansion, analyzing its economic significance and some of the problems it faced in its expansion.

Wool produced in estancias and farms in the province of Buenos Aires found its main consumers in the European market, particularly among French and Belgian manufacturers. Chapter 6 turns to the international market for wool, the role of Argentine produce therein, and the competition it found from wool grown in other areas. The trend and fluctuations of prices paid for River Plate wool in Buenos Aires, Antwerp, and Le Havre between the 1850s and 1880s show striking similarities, revealing the close relationship between those markets. Although the significance of Argentine produce in certain areas affected the course of prices in the corresponding markets, in the long run international demand for wool and the overall situation in the consuming areas determined those prices. Throughout the century, periods of crisis followed periods of expansion, and producers in Buenos Aires found their own way of responding to such fluctuations.

The link between producers and the market was provided by a variety of commercial networks, which are also described in Chapter 6. The persistence of old trading methods and the development of new channels to deal with wool resulted in a complex system in which local and foreign houses and agents shared a profitable but risky activity. Declining costs of commercialization and more efficient services contributed to bring a higher degree of certainty to the wool trade in the 1870s. However, the problem of who profited from these improvements depended entirely on the relationship established between wool growers and trade networks, and on the relative power of each.

The difficulties in consolidating a stable and efficient money market were among the main problems found in the development of a capitalist structure of production and circulation in Buenos Aires. The expansion of the pastoral industry by mid-century was to exert increasing demands on the primitive financial networks that were in operation during the first decades of the century. Until the 1850s, specialized agencies were almost nonexistent, and the commercial networks covered most of the services that were later to be provided by banks.

In the following decades, this situation was only partially solved by the establishment of several banks and agencies in Buenos Aires. Chapter 7 describes the main capital requirements of the pastoral sector and analyzes the way in which these were met. It concentrates particularly on the provision of credit through banks, commercial houses, and other more informal institutions. Short- and medium-term credit prevailed upon long-term schemes, and money was relatively scarce and expensive to wool growers in the period under study. However, not all producers were equally placed in relation to financial networks, and large estancieros generally obtained better arrangements than small wool growers and farmers.

At the back of these descriptions of the different aspects of sheep raising and the wool trade, the guiding question was how capitalist organization, reproduction, and accumulation were achieved within the pastoral sector in the period that witnessed such an impressive process of growth. That is why I chose to analyze only certain aspects of the history of the pastoral industry in Argentina—a vast subject which allows for different approaches and interpretations. Thus, this study emphasizes the way in which land and labor markets were formed during the period, how rural enterprises were organized and realized their profits and investments, and how the commercial and financial networks operated in relation to the wool trade. However, only secondary attention is paid to other aspects of the problem which should be studied in order to have a more complete interpretation of the development of the pastoral industry and of the process of accumulation within the area. For example, the role of the state in this development as well as the relationship between the pastoral sector and the rest of the economy clearly deserve more thought and consideration than I have been able to give them. In this sense, I hope that my work will stimulate further research on those, as well as on many other aspects of this neglected subject of Argentine history.

Notes

1. The best accounts of the development of the pastoral sector are to be found in José G. Chiaramonte, *Nacionalismo y liberalismo económicos en Argentina, 1860–1880* (Buenos Aires, 1971); Horacio Giberti, *Historia económica de la ganadería argentina* (Buenos Aires, 1961); Ricardo Ortiz, *Historia económica de la Argentina* (2 vols., Buenos Aires, 1964), vol. 1.

2. José Ingenieros, *Sociología argentina* (Madrid, 1913); Jonathan Brown, *A Socioeconomic History of Argentina, 1776–1860* (Cambridge, 1979); Giberti, op. cit.; Ortiz, op. cit.

3. Aldo Ferrer, *La economía argentina* (Buenos Aires, 1963); Haydee

Gorostegui de Torres, *Argentina, la organización nacional* (Buenos Aires, 1972); Tulio Halperin D., *Historia contemporánea de América Latina* (Madrid, 1969) and *Argentina, de la Revolución de Independencia a la Confederación Rosista* (Buenos Aires, 1972). More recently the same point has been made by Alfredo Pucciarelli, *El capitalismo agrario pampeano, 1880–1930* (Buenos Aires, 1986).

4. This perspective is adopted, among others, by Roberto Cortés Conde, *El progreso argentino, 1880–1914* (Buenos Aires, 1979), and Carlos Díaz Alejandro, *Ensayos sobre la historia económica argentina* (Buenos Aires, 1975).

5. This perspective is adopted by Ferrer, Giberti, and Ortiz, among others; see notes 1 and 3.

6. André Gunder Frank, *Capitalism and Underdevelopment in Latin America* (New York, 1967); Fernando H. Cardoso and Enzo Faletto, *Dependencia y desarrollo en América Latina* (Mexico, 1969), and also their "Postscriptum a *Dependencia y desarrollo en América Latina*," *Desarrollo Económico*, 17, n. 66 (July–Sept., 1977). A large number of works deal with the subject of dependency; for a very good review and critique of the whole debate, see Gabriel Palma, "Dependency: A Formal Theory of Underdevelopment or a Methodology for the Analysis of Concrete Situations of Underdevelopment?" *World Development*, 6, 881–924.

7. Palma, op. cit.

8. Ernesto Laclau, "Modos de producción, sistemas económicos y población excedente. Aproximación histórica a los casos argentino y chileno," *Revista Latinoamericana de Sociología*, 5 (1969), 276 ff.; Jorge F. Sábato, *La clase dominante en la Argentina moderna. Formación y características* (Buenos Aires, 1988) (first published in 1979).

9. Laclau, op. cit., 296.

10. The hypothesis of the existence of this international Ricardian rent in favor of Argentine produce is not easy to prove in the specific case of wool. Without attempting to arrive at final conclusions, however, it is possible to point to several facts which seem to corroborate that, at least during certain periods, there must have been such a transference. The relatively high fertility of the Buenos Aires soil becomes evident by comparing the number of sheep that could be raised per hectare in Buenos Aires with that of other wool-producing areas, which even after fencing did not reach the high proportions of the unfenced runs of the River Plate plains. In the early days of expansion, this advantage undoubtedly accounted for the higher profits made by Buenos Aires sheep raisers when compared to Australian ones, who—in spite of paying less for labor and receiving better prices for their staples—realized lower profits than did their Argentine counterparts. See among others, G.

Abbott, *The Pastoral Age: A Re-examination* (Melbourne and Sidney, 1971), 89–107; Francisco Seeber, *Importance économique et financière de la République Argentine* (Buenos Aires, 1888), 108–9.

11. There are a great number of works on the period 1880–1930. See, for example, the following bibliographies: Sergio Bagú, *Argentina 1875–1975. Población, economía, sociedad. Estudio temático y bibliográfico* (México City, 1978); Tulio Halperin Donghi, "Argentina," in R. Cortés Conde and S. Stein (eds.), *Latin America: A Guide to Economic History 1830–1930* (Berkeley, 1977).

12. Although this view has been already questioned by specific works on the merchant community of Buenos Aires (particularly for the first half of the nineteenth century), it is still upheld by recent studies, such as Pucciarelli, op. cit. For a debate on the more general problem of foreign control over trade and finance in Latin America, see *Latin American Research Review*, 21, No. 3 (1986), 145–56.

13. For 1854, *Primer censo de la República Argentina, 1869* (Buenos Aires, 1872), 19. For 1879 and 1881, *Censo general de la provincia de Buenos Aires, 1881* (Buenos Aires, 1883), 124.

14. The chosen counties are the following: Arrecifes, Baradero, Brandzen, Cañuelas, Carmen de Areco, Chacabuco, Chascomús, Chivilcoy, Ensenada, Exaltación de la Cruz, Las Heras, Lobos, Luján, Magdalena/Rivadavia, Mercedes, Monte, Navarro, Pergamino, Pilar, Ramallo, Ranchos, Rojas, Salto, San Andrés de Giles, San Antonio de Areco, San Nicolás, San Pedro, San Vicente, Suipacha, Zárate.

15. See Tulio Halperin Donghi, "Canción de otoño en primavera: previsiones sobre la crisis de la agricultura cerealera argentina (1894–1930)," in *Desarrollo Económico*, 24, no. 95 (Oct.–Dec., 1984).

16. See, for example, Cortés Conde, op. cit.

17. See, for example, Ferrer, op. cit.; Giberti, op. cit.; and Jacinto Oddone, *La burguesía terrateniente argentina* (Buenos Aires, 1930). This perspective has inspired a number of more recent works that, while asserting the capitalistic nature of Argentine economy at the turn of the century, underline the importance of international differential rent and point to the landowning oligarchy as the main local recipient of the surplus thus transferred to the River Plate. See, for example, Laclau, op. cit.; Guillermo Flichman, *La renta del suelo y el desarrollo argentino* (Mexico, 1977), and Pucciarelli, op. cit.

18. The pragmatic perspective is clearly embraced by Roberto Cortés Conde, op. cit. For the critical perspective, see, among others, Ofelia Pianetto, "Mercado de trabajo y acción sindical en la Argentina, 1890–1922," *Desarrollo Económico*, 24, No. 94 (July–Sept., 1984); Ernesto Laclau, op. cit.; and Hilda Sabato, "La formación del mercado de trabajo en Buenos Aires, 1850–1880," *Desarrollo Económico*, 25, No. 96 (Jan.–March, 1985).

1

Sheep Raising in Buenos Aires Province

Outside the agricultural farms of Buenos Aires the great
sheep-walks almost monopolise the campo.
—Wilfrid Latham, *The States of the River Plate,* 1868.

By the late 1840s Buenos Aires province was predominantly a
cattle-raising area, with a narrowly diversified economic structure.
Three million head of cattle constituted the main resource of the
province. These animals were of inferior quality, having grazed
on the natural grasses of the pampas, and kept in open fields
under the care of one man who tended at least three thousand
or more head. These cattle were just good enough to produce
what was required of them—hides and jerked beef, the staple
exports of the province. These were traditional Argentine products, with the former being sent mainly to Europe, and the latter
sold to feed the slave populations of Brazil and Cuba. This primitive picture was hardly modified by the innovations introduced
by mid-century, which were directed toward improving the quality of the meat.

However, a decisive change was to be brought about not by
cattle, but by sheep. It would be through the export of wool that
Argentina would gain a significant place in the world market,
thus developing its internal productive capacity and promoting
a rapid process of capital accumulation centered in Buenos Aires.
Throughout this book this transformation, as it affected different
dimensions of the agrarian structure, will be portrayed. In this
chapter, the process of expansion of the pastoral industry in Buenos Aires, its main characteristics, limitations, and problems, will
be described.

The First Experiments in Sheep Breeding

In 1810 the country had a stock of two to three million sheep, but
these were of extremely poor quality and occupied only marginal

lands. During the following three decades, cattle raising on an extensive scale was the only important productive activity carried out in the rural areas of Buenos Aires. Abundance of land, scarcity of labor and capital, and a secure market, though not a very dynamic one, were some of the factors the elites of the province took into account when supporting the development of cattle raising to complement the financial and commercial activities of the capital city and port.

Yet by the late 1820s and 1830s a pioneer effort would be made in relation to the pastoral industry, as a few estancieros—most of them foreign-born—imported pure breeds from Europe in order to experiment in crossbreeding them with the *criollo* (or native) flocks. For foreigners who had made a fortune in commercial or financial activities and sought to invest in a productive enterprise, sheep raising offered good opportunities, as it required less initial investment than cattle, and was not monopolized by native estancieros, as the cattle trade was. Although commercial networks for the produce of sheep were not yet organized, these pioneer entrepreneurs could take advantage of their connections abroad to acquire the pure breeds, and to sell the wool. Very soon, they would find excellent opportunities for expansion. The international market was opening up to wool from faraway lands, as the traditional sheep-raising areas were turning to other products, leaving an expanding textile industry without the necessary raw material. Australia, South Africa, and Argentina's River Plate would then become the new suppliers of this product, which was increasingly demanded by manufacturers in England, France, Germany, Belgium, and the United States.

In 1822 wool represented only 0.94% of the exports of the province, but this figure went up to 7.6% in 1836 and 10.3% in 1851.[1] *Cabañas*, like those of C. Stegmann, P. Plomer, J. Harratt and P. Sheridan, were in full swing by the end of the 1840s, and in many areas sheep were gradually displacing cattle from the prominent place they had held for decades.[2] MacCann would observe, "For twenty leagues around the city of Buenos Aires the estancias, or cattle farms, ought rather to be called sheep farms."[3]

The new industry had to face many problems. In spite of the reproduction of the herds, most of the four million sheep were still inferior, their wool being rejected by international buyers who pressed for better quality. Furthermore, the scarcity of labor power, and the absence of an organized network to meet the different needs of the trade (such as technical assistance, or the periodic visit of shearing squads) added to problems in the commercialization of the products, which was hampered by blockades of the port of Buenos Aires and by political conflicts and struggles. Natural factors, such as drought and flood, wild dogs and *vizcachas*, thistle and big burr, were still other problems to be faced by those pioneer estancieros interested in sheep raising.

In order to cope with the ups and downs of the international wool market, in the 1840s the breeders started to industrialize tallow for export purposes.[4] The first plant to that end was built in 1842, and many others would be set up later in different towns throughout the province, every time the wool market showed signs of crisis. Meat also began to be commercialized for local consumption, and although it would prove hard to change the eating habits of the population, mutton was to become a common dish on the table of the rural workers of the 1860s and 1870s. The Rambouillet sheep became the favorite breed, combining as it did, good quality of wool with a corpulent body.

First Crisis and the Great Expansion

During the 1850s the industry continued to expand, and by the end of the decade fourteen million sheep were in stock. Cross-breeding had proceeded very quickly indeed, and in most of the northern counties by 1860 the clip results show a clear predominance of *mestizo* wools.[5] As we shall see in later chapters, new hands were available for the jobs arising from sheep raising, technical improvements were being introduced, and commercial networks were developing to deal with the new exports. In times of expansion the business would prove very profitable, requiring little initial investment, and rewarding investors with quick returns.

But the pastoral industry was soon to start feeling the consequences of its tight links with the international market, and the repercussion of European crises in its internal development. The first shock was felt in 1857–58, after the short expansion of demand caused by the withdrawal of Russia from the market as a result of the Crimean War. The price of wool went up, and the River Plate growers responded by expanding production. They were soon confronted with a drastic drop in prices, as Russia returned to its role of supplier, and demand fell far short of existing stocks.

This pattern would become recurrent in the history of the Buenos Aires pastoral industry. The response of the local growers generally consisted of two successive steps, adopted in order to palliate the consequences of the crises. First, during the critical years the tendency was to export more wool in order to compensate for the drop in prices; second, more animals than usual were killed and used for tallow and sheepskins, thus diminishing the flocks for the year to come, as a method of controlling production. Thus, by 1858, exports of wool had increased while the following two years saw an expansion in the export of sheepskins.[6]

The immediate result of these measures was a new cycle of expansion which reached its peak in 1865 in what could be called the boom of the pastoral industry. Exports swelled, herds multiplied, production boomed. Antwerp increased demand, and Belgium became by far the largest buyer of River Plate wool. Forty million head of sheep now grazed on the prairies of the province, a herd that had increased at the peak rate of 23.36% per year during the quinquennium. The wool exported went up from 12 million kg in 1859 to over 48 million kg six years later, when wool had undoubtedly become the main export of the province, and of the country as a whole. Thus, although cattle raising still constituted a widespread activity in Buenos Aires province, it was surpassed in importance by sheep raising; cattle were also physically displaced to marginal areas and used as *animal de avanzada* in fields that required 20 to 30 years of preparation before sheep could graze on them.

A number of factors contributed to this sheep boom, and here I will only mention them briefly, as most will be dealt with in later chapters. Although the prices for wool followed a declining trend during the five years before 1865 both in Buenos Aires province and in the importing countries, demand from all the main countries increased steadily as did the total income received from the wool sales.[7] Internally, the commercialization and financing networks were running smoothly by this period, and transportation was developing rapidly, reducing freights and losses. Irish, Basques, French, and Scots had become familiar faces in rural Buenos Aires province, providing the necessary and adequate labor force for the expanding industry. In some cases, these immigrants developed into farmers, investing their small capital in the business, which asked for relatively little initial investment but required skilled labor power, frequently provided by the family of the farmer.

Profits were high, and capital was flowing into the business. On the one hand, wool production and sheep raising attracted private investments from both estancieros who were already engaged in raising livestock and from newcomers to the rural enterprises who had some capital to invest and who chose this promising field. On the other hand, the interests of the provincial government brought about an inflow of increasing amounts of social capital in the form of roads, railways, wars against the Indians, and other similar enterprises.

Official action favored the pastoral industry in several different ways. Although in 1862 export duties on wool and sheepskins went up from 5% to 10%, other measures counterbalanced this.[8] For example, up to 1864 the government kept a devalued currency, which favored all exporting sectors. More specifically, it encouraged the holding of exhibitions and the introduction of pure breeds by giving legal and financial assistance for those purposes. Also, the *Código Rural* was passed in this period, thus contributing to the legal organization of the countryside.

Sheep estancieros, for their part, had also started to organize, acting as a pressure group on different occasions, and finally forming the core group of the *Sociedad Rural* created in 1866.

Optimism and high expectations were the prevalent features of the day. Prices of land, labor force, and sheep all went up in the first years of the decade. But soon this optimism would prove excessive, as the somber years of a new cyclical crisis approached, which would hit the pastoral industry more seriously than ever before.

An Unparalleled Crisis

The pastoral industry was probably the sector most severely hit by the general crisis of the mid-1860s, a blow from which it would not recover until the following decade. The shortage of money coupled with the valorization of the Argentine peso in 1865, reversing the policy of previous years, was undoubtedly unfavorable to the export sectors, the *ganaderos* among them. The consequent rise in the rates of interest brought about speculation, which reached its highest point during the shearing season, when sheep owners were in need of additional capital to finance their operation and had to pay up to 30% annual interest on the money they borrowed.[9]

These local monetary problems were aggravated by an international commercial crisis resulting from overproduction. Prices of wool dropped from 3.14$oro per 10 kg in 1864 to 2.65$oro in 1867, to go down even further in the following years, reaching in 1869 the lowest point of the period between 1855 and 1890, quoted at 2.19$oro per 10 kg.[10] The rate of increase in exports also fell, although there would be an increase in the absolute quantities of wool exported until 1870, when this trend was reversed for a couple of years. The situation was further troubled in 1867 by the protectionist law passed in the United States, heavily taxing imports of unwashed wool, which constituted the bulk of Argentine exports to that country.

The internal situation of the industry was not encouraging, either. The great increase in the flocks during the previous years had not been met with an equivalent expansion of the frontier. On the contrary, after the Indian raids of the late 1850s, little had been done to protect the affected areas, and by the mid-1860s the war with Paraguay prevented any diversion of troops from the main front. The Law of 1864 regarding public lands was also seen by the rural sectors as an impediment to new settlements. Therefore, overstocking ensued.[11] Prices of sheep declined, and many were sold to *saladeros* and tallow plants, as a means of getting rid of surplus stock. Hides and tallow were favored in this period, as there was great demand for them due to the proliferation of armed conflicts both at home and abroad.[12]

The war with Paraguay affected both the supply and the price of labor force, as native rural workers were often taken for the front. However, the general effects of the war seem not to have been unfavorable for the economy of Buenos Aires province, as large amounts of Brazilian gold were introduced, and supplying the army became big business.[13]

This unfavorable period for the pastoral industry led sheep estancieros to defend their interests by consolidating their organization as a pressure group and by creating the Sociedad Rural in 1866. Many of them became critical of the economic liberalism that had reigned supreme in the country since the early 1850s, and they began to propose protectionist measures to foster the development of agriculture and manufacturing and to promote the diversification of the productive structure. At one point they even proposed to establish a textile factory, which was to include well-known estancieros on its governing board.[14]

The estancieros also pressed the government in matters related to their particular interest, such as the export duties on wool and sheepskins, which were reduced from 10% to 8% in 1866, and then to 6% in 1868, to drop finally to 2% by 1870. In monetary policy, they favored the establishment of the *Oficina de Cambios*, to guarantee an adequate supply of paper money, which had been scarce since printing had been suspended in 1861.[15]

Thus by the end of the decade the situation had started to improve for the exporting sectors, but nevertheless the optimistic days of the early 1860s were gone.

A Period of Moderate Expansion

The 1870s and early 1880s proved favorable to the pastoral industry, though there were ups and downs caused by the vulnerable situation of the province as a supplier in the world market, and also by local problems affecting the sector.

By 1871–1872 an increase in the international price of wool once again acted as a stimulus for the expansion of its production in Buenos Aires province, while tallow and sheepskins declined. Thus, the short-lived peak of demand caused by the end of the Franco-Prussian War led growers to flood the market, only to feel the contraction that soon followed.[16] But a more serious situation arose the following year when a new international crisis broke out, followed locally by a severe depression which lasted for several years.

Expansion of credit, inflow of capital through international loans, rise in imports, unfavorable balance of trade, and speculation on a large scale were the features that had prevailed in the years previous to 1873 in Argentina. Collapse followed, with contraction of credit and money supply, bankruptcies, and usury. The price of local goods in the international market had dropped, and only a policy of austerity on the part of the government could stop the outflow of gold and the continuation of speculation. That is what President Avellaneda did, amid controversial debates regarding the best way out of the crisis.

This time the depression hit the commercial and financial interests harder than any others, and the ganaderos could be said to have coped quite well with the situation. Except in the worst years of 1873 and 1874, there was a relative expansion both in the export of wool and in the income this represented during the decade (Table 1, Figure 1). Sheep stocks, however, did not increase, as a serious epidemic in 1873–1874 and floods in 1877 killed a large number of animals.[17] Improved technical conditions nevertheless allowed for better clips than in previous decades, and the production of wool continued to expand in spite of the relative stagnation, and even decline, of the flocks (Figure 2).

Shortage of labor force and of land were the main complaints put forward by sheep estancieros during this period, but there were other, favorable factors to compensate for those problems. Thus, although most of the credit had been channeled toward speculation, the Banco Provincia had actually increased its loans to rural entrepreneurs, who used the money mainly to improve the technical aspects of their estates.[18] Railways were expanding, and so were commercial networks; capital continued to flow into the area, and the number of farms and estancias kept increasing. By the end of the decade, international prices of wool would go up again, land available would expand after the campaigns carried out by Alsina and Roca, and the security of the frontier would be guaranteed by the extermination of the Indians. Thus, the 1880s started in a favorable mood, the prices of wool rising, exports also in the upward trend, and production increasing.

Sheep raising was indisputably the main productive activity carried out in the province, and its products had become the staple exports of the country as a whole (see Table 1). After three decades of development, the pastoral industry had reached a mature state,

Table 1
Main Argentine Exports, 1865–1882

Year	Wool tons	Wool val.[a]	Salted meat tons	Salted meat val.	Tallow & grease tons	Tallow & grease val.	Hides units	Hides val.	Sheepskins tons	Sheepskins val.	Horsehides units	Horsehides val.	Total hides & skins (val.)	Total exports (val.)
1865	54,926	12,246	35,539	1,217	16,635	2,269	2,031	5,730	8,088	1,232	185,600	256	7,218	26,491
1866	54,014	12,275	22,893	974	14,720	2,200	1,980	5,413	10,396	1,487	126,500	174	7,074	25,878
1867	53,421	14,574	31,980	1,039	27,633	4,130	2,399	7,051	13,774	1,993	115,600	153	9,197	32,125
1868	70,230	12,241	27,774	945	37,718	5,311	2,355	7,436	14,781	1,724	113,000	171	9,331	31,800
1869	72,451	10,708	37,905	1,284	54,094	7,610	2,928	11,620	21,718	1,987	150,400	179	13,786	34,995
1870	55,701	6,861	38,730	1,255	47,540	6,673	2,712	8,289	26,407	1,850	102,100	154	10,293	29,248
1871	71,565	7,471	32,238	1,060	34,281	4,527	2,433	7,395	20,854	1,389	120,400	173	8,957	26,426
1872	92,426	16,352	41,659	2,111	53,355	7,385	3,240	10,749	33,177	4,159	208,500	337	15,245	45,743
1873	83,733	19,605	30,813	1,383	40,236	5,487	2,671	9,680	25,175	4,280	149,400	279	14,239	45,869
1874	80,207	17,967	25,435	1,009	15,107	2,071	3,106	11,755	24,503	4,303	255,300	442	16,500	43,405
1875	90,724	19,960	34,048	1,363	33,472	4,677	2,883	11,743	29,525	5,164	242,500	453	17,360	50,331
1876	89,546	19,680	29,543	2,016	37,433	5,641	2,325	7,943	27,598	4,845	195,900	318	13,106	46,539
1877	97,344	18,112	38,719	2,712	27,431	4,134	2,488	7,225	27,854	3,934	262,200	478	11,637	43,326
1878	81,894	14,754	33,579	2,364	21,110	3,179	2,239	6,430	27,873	3,904	202,000	367	10,701	36,313
1879	92,112	21,674	32,310	2,812	15,538	2,031	2,337	8,150	25,089	3,966	217,400	292	12,408	47,765
1880	97,518	26,754	26,109	2,978	11,945	1,759	2,791	10,898	29,079	5,280	326,900	461	16,639	56,497
1881	104,757	30,739	22,399	2,545	10,687	1,428	2,192	8,844	22,342	4,491	280,600	405	13,740	56,069
1882	111,095	29,033	26,997	3,756	18,434	2,699	1,945	8,286	22,358	4,095	213,000	416	12,797	58,441

[a]Value expressed throughout in thousands of *pesos fuertes*.

Source: José C. Chiaramonte, *Nacionalismo y liberalismo económicos en Argentina, 1860–1880* (Buenos Aires, 1971), 38–39.

Figure 1

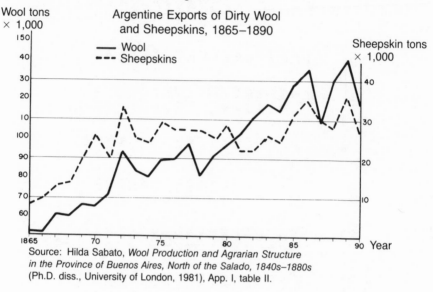

Argentine Exports of Dirty Wool
and Sheepskins, 1865–1890

Source: Hilda Sabato, *Wool Production and Agrarian Structure
in the Province of Buenos Aires, North of the Salado, 1840s–1880s*
(Ph.D. diss., University of London, 1981), App. I, table II.

Figure 2

Sheep Stock in the Province
of Buenos Aires, 1850–1910

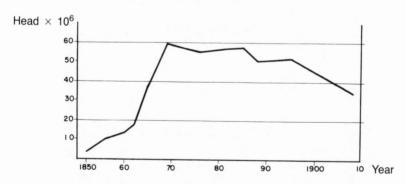

Source: Sabato, op. cit., App. I, table I.

with an organization of production, commercialization, and finance adequate to its needs. In spite of these achievements, however, its structural situation was still quite vulnerable. Almost total dependence on the international market, plus subordination to natural phenomena in spite of technical improvements that had been introduced, were to hit the business continuously, making it rather hazardous and unstable.

Prelude to Decline

Although during the rest of the century wool would remain the staple export of the province and of the whole country, important changes were to take place in the agrarian structure during the late 1880s and the 1890s, bringing about a transformation for rural Buenos Aires province. This work will not deal with this stage in the development of the productive structure, but I shall briefly point out the main factors leading to the decline of wool production and sheep raising in the province.

New requirements in the international market, the expansion of the urban population in Argentina, and the problems periodically faced by a productive structure that relied heavily on the export of a single product were to bring about a gradual change in the pastoral industry. An increasing interest in meat would lead first to the export of livestock, but soon afterwards to experiments in freezing and chilling, and the establishment of the first packing plants. In 1880 no mutton was exported, but seven years later one million head were sent by four different firms to France and England. The River Plate Fresh Meat Co. was the first to develop this trade, exporting over 120,000 carcasses in 1884.[19]

Sheep were at first chosen for this experiment, as decades of crossbreeding had provided most of the flocks with the adequate conditions for consumption as mutton. But merinos did not provide the best type of meat for freezing, and therefore a gradual process of new crossbreeding would start with the introduction of the Lincoln breed. The transformation was soon to be felt, and Gibson has described the process as follows:

Not until 1882 did the Lincoln become a generally popular breed, and today it disputes the land with the merino in every corner of

the Republic. This change has been brought about by three causes.
. . . First, a succession of wet seasons, commencing in 1877 and
continuing with few gaps until 1884, had occasioned heavy losses
in the merino stock, particularly upon those lands near the seafront.
. . . Second, the frozen meat trade . . . soon assumed titanic proportions. . . . Finally, in 1884 a fall in merino wools became accentuated, and long wools, especially *cross wools*, sold at better prices.
. . . The result has been a rush after Lincolns for crossing purposes.[20]

Merinos would soon be displaced to other areas of the country,
mainly Entre Ríos, Corrientes, La Pampa, and Patagonia.

The hitherto undisputed leading place of wool among the exports of the country became questioned. Not only had new products begun to compete for first place, but also though production
and stocks were relatively constant during the 1880s (with oscillations), toward the end of the decade and in the early 1890s there
was a drastic drop in the income from this export, as prices fell
in the international market. The final blow would come by the
end of the century, when sheep breeding in Buenos Aires lost its
place to cattle raising and agriculture.

Sheep on the Move

By 1858 the territory under control of the province of Buenos
Aires was about twice as large as the area north of the Salado
River, and it was extended even more in 1877 to include most of
its present territory. Nonetheless, the counties of the north were
in a privileged position throughout the period under consideration. Most of them were close to the capital city, and they had
benefited from that situation by experiencing an early process of
development. Thus, the first rural activities carried out in the
territory of the province were located in this area, at first in the
immediate vicinity of Buenos Aires, but later on spreading to the
south, north, and west. Only the frontier counties of the extreme
west had to wait until the 1860s to share that situation (see Maps
6, 7, and 8).

Sheep breeding was no exception, and it was initially developed
in this region of the province. The southern counties of Cañuelas,
Las Heras, San Vicente, and Ranchos were the first districts where

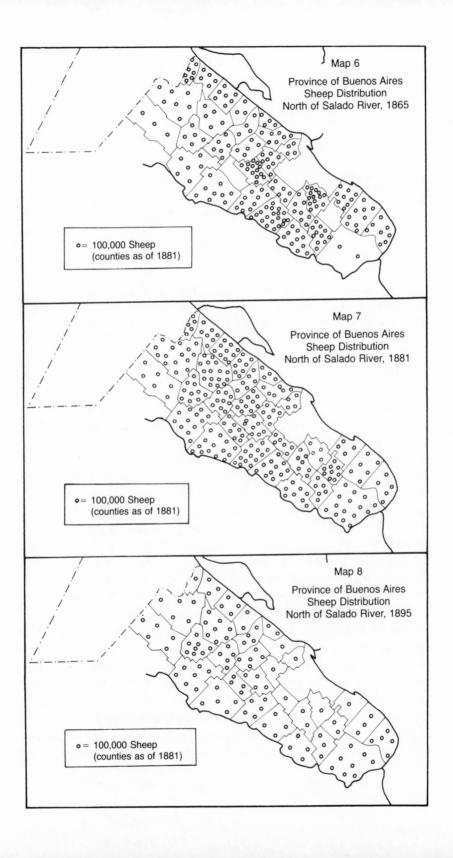

Map 6
Province of Buenos Aires
Sheep Distribution
North of Salado River, 1865

o = 100,000 Sheep
(counties as of 1881)

Map 7
Province of Buenos Aires
Sheep Distribution
North of Salado River, 1881

o = 100,000 Sheep
(counties as of 1881)

Map 8
Province of Buenos Aires
Sheep Distribution
North of Salado River, 1895

o = 100,000 Sheep
(counties as of 1881)

Table 2
Sheep and Cattle in the Province of Buenos Aires,
Total Province and by Regions, 1865, 1881, 1888

	Sheep (thousand head)			Cattle (thousand head)		
Year	North[a]	Center & South[b]	Total	North[a]	Center & South[b]	Total
1865	23,246	14,736	37,982	1,519	4,619	6,138
1881	27,147	30,691	57,838	1,440	3,314	4,754
1888	17,100	34,139	51,239	1,534	6,809	8,343

[a]Includes all counties north of the Salado River.
[b]Includes all counties south of the Salado River.
Sources:
1865: Rep. Argentina, *Registro Estadístico de la República Argentina, 1865*, 246–247.
1881: Pcia. de Buenos Aires, *Censo General de la Pcia. de Buenos Aires, 1881*, 338–339.
1888: Pcia. de Buenos Aires, *Censo Agrícolo-pecuario de la Pcia. de Buenos Aires, 1888*, 240–357.

breeding experiments took place, in famous cabañas like those of Plomer, Sheridan (Los Galpones), and Harratt (Los Galpones Chicos).[21] Soon, Chascomús, Lobos, Navarro, and Monte were incorporated into this leading sheep-raising area, and by the early 1860s production had so expanded that most of the counties north of the Salado River were also becoming pastoral. By 1865, the flocks in this area represented 58% of the provincial total, but in the 1870s, with the movement of the frontier and the development of the railways, sheep raising expanded southward. And although by 1881 the northern counties still show higher densities than those of the south, their share in the total provincial stock had dropped to 45%, falling even further in 1888 to 32% (see Table 2).

While the south was becoming increasingly pastoral, the northern districts were starting to diversify their structure of production as a response to several different factors. First of all, I should point out that, although our emphasis here has been on the expansion of sheep raising as the activity that prevailed in the area under study, in no way should we think of it as a complete and uniform sheep run. Both cattle raising and agriculture were secondary activities during the period under consideration, but in certain counties they experienced an early process of expansion. Such was the case of Chivilcoy and Baradero, for example, where agricultural colonies had been established in the late 1850s and 1860s, and of many of the counties close to the capital, which had produced farm goods and vegetables for the consumption of the expanding city ever since colonial times. At the same time, although cattle had been displaced by sheep in this area, most of the large estancieros kept some herds on their estates for the production of meat, hides, and grease. Thus, by 1881, 409,000 hectares out of 58 million were under cultivation north of the Salado, and 1.4 million head of cattle shared the prairies with 27 million sheep.[22]

By the late 1880s a different process was also taking place. I have mentioned how an increasing demand for meat from both the local and the international markets had led to experiments in chilling and freezing, and to the *desmerinización* of the flocks, in order to adapt them to the taste of mutton consumers in Europe. Very soon, however, beef was also being exported, following a crossbreeding effort by some *cabañeros* and estancieros who saw a promising field in the newly opened trade. By the end of the century cattle were replacing sheep, and with them came agriculture on a large scale, as a system of rotation of crops was introduced, whereby a plot of land was cultivated by tenants who, after two years of growing grain and linseed, left the land in the third year planted with alfalfa, for the landowners to feed their cattle.[23] The development of the railway network and the entry of an immigrant labor force also contributed to the development of this new type of production.

Sheep were displaced to other areas where the soil was not so productive, but where land was cheaper, and railways made the extraction of products feasible. The Indian menace had disappeared after the *Campaña del Desierto* and the new territories were being incorporated into productive use. Thus, while counties like Tandil were now devoted to sheep breeding, those counties closer to Buenos Aires, like Pilar, Lujan, San Vicente, Cañuelas, and others with equivalent natural and geographical conditions, were shifting to intensive land use, becoming agricultural and farming areas par excellence. The rest of the counties started to produce grain and cattle for export purposes, and although sheep did not completely disappear from this region, by the beginning of the twentieth century they were being increasingly pushed southward.

Notes

1. José Chiaramonte, *Nacionalismo y liberalismo económicos en Argentina (1860–1880)* (Buenos Aires, 1971), 33.
2. See Herbert Gibson, *The History and Present State of the Sheep Breeding Industry in the Argentine Republic* (Buenos Aires, 1893), and Estanislao Zeballos, *Descripción amena de la República Argentina* (3 vols., Buenos Aires, 1881/88), vol. 3.
3. William MacCann, *Two Thousand Miles' Ride through the Argentine Provinces* (2 vols., London, 1853), vol. 1, 149.
4. See Gibson, op. cit., 72; Zeballos, op. cit., 45–49; Emile Daireaux, *Vida y costumbres en el Plata* (2 vols., Buenos Aires, 1888), vol. 2.
5. By 1865 over 90% of the pastoral stock in the province was crossbred. See *Registro Estadístico de la República Argentina* (1865).
6. *Parliamentary Papers, Commercial Reports,* (b) Embassy and Legation, 1876, vol. LXXIII, 206–7.
7. Hilda Sabato, *Wool Production and Agrarian Structure in the Province of Buenos Aires, North of the Salado, 1840s–1880s* (Ph.D. dissertation, University of London, 1981), App. I, Tables II & III.
8. Chiaramonte, op. cit., 88–90.
9. Chiaramonte, op. cit., chap. 2.
10. See below, table 20. Throughout the text, I have used the *peso oro* as the main monetary unit. Although it was established by a law of 1881, I have used it as a *moneda de cuenta* for the whole period, in order to facilitate comparisons. Conversions from paper pesos and pesos fuertes have been made according to equivalences included in table 22.
11. For accounts of these problems, see, for example, *The Brazil and River Plate Mail,* issues of 7 Dec. 1863, 21 June 1864, 7 Sept. 1864; Eduardo Olivera, "Nuestra industria rural en 1866," in *Miscelánea, Escritos económicos, administrativos, económico-rurales, agrícolas, ganaderos, etc.* (2 vols., Buenos Aires, 1910), vol. 1, 62.
12. *The Brazil and River Plate Mail,* 22 Oct. 1867; *Le Courier de la Plata,* 8 July 1867.
13. See Chiaramonte, op. cit., 66; Francisco Latzina, "El comercio argentino antaño y hogaño," in *Censo Agropecuario Nacional. La ganadería y la agricultura en 1908,* Monografías (Buenos Aires, 1909), 577.

14. For a full account of the protectionist movement, see Chiaramonte, op. cit. On the Sociedad Rural, see Tulio Halperin Donghi, *José Hernández y sus mundos* (Buenos Aires, 1985), chap. 6.

15. H. Cuccorese, *Historia de la conversión del papel moneda en Buenos Aires (1861–67)* (La Plata, 1959).

16. *The Brazil and River Plate Mail*, 7 Oct. 1871; *L'Economiste Français*, 1873, 73, and 1877, 275, 338, 754.

17. *L'Economiste Français*, 1875, 523.

18. Chiaramonte, op. cit., 239.

19. The other three firms were Sansinena, Terrason, and Nelson (Zeballos, op. cit., chap. 13).

20. Gibson, op. cit., 37–38.

21. See Gibson, op. cit., 200 ff.; Zeballos, op. cit., vol. 2, chaps. 5, 6, and 7; John Murray, *The Story of the Irish in Argentina* (New York, 1919), chaps. 12 and 13.

22. Provincia de Buenos Aires, *Censo General de la Provincia de Buenos Aires*, 1881 (Buenos Aires, 1883), 306, 338.

23. For a detailed description of this system, see James Scobie, *Revolución en las pampas. Historia social del trigo argentino, 1860–1910* (Buenos Aires, 1968).

2

Land Tenure and Distribution

Le sol est la merchandise la plus abondante en ce pays;
lorsque sa valeur augmente c'est un enrichissement
général.
—*Le Courier de la Plata*, 31 December 1880.

The Rosas Era

Land was perhaps the most abundant resource in Buenos Aires
province of the 1830s and 1840s, and although much of the territory that today belongs to the province was controlled by Indians, large extensions of land within the frontier were still scarcely
occupied by the white man. In 1830 more than half of the present
territory had no legal occupants, whether landowners or "enfiteutas."[1]

After the Revolution of 1810 cattle raising expanded throughout
the province, and the rural areas gradually came under the hegemony of the estancieros. The alliance of this ascendant class
with the urban government elite was to bring about—particularly
after the 1820s—an aggressive policy of expansion of the frontier,
and a new approach to private property.[2] The secondary place
that had been reserved in the colonial order to rural interests, as
well as the availability of land to almost anyone who wanted to
make use of it, had contributed to postpone the problem of distribution of that important means of production. At a time when
the concepts of property and appropriation of land were not necessarily coincident, landowners, tenants and squatters shared the
benefits of open fields where cattle could roam endlessly, and
where no fences could stop their search for better grasses, fresh
water or shady trees. Yet this situation proved more and more
inconvenient for the larger estancieros, who wished to ensure

41

their rights to a tract of land, the possibilities to expand and improve it, and the elimination of competition.

Expansion of cattle raising, and a shrinking frontier neglected by a government too busy with the war of independence, called for a new policy towards land. Thus, in order to foster the settlement of population in frontier areas, as early as 1817 the *Directorio* decided to give away public lands in those areas, and soon afterwards the provincial government launched a series of campaigns to gain control of territories then subject to Indian raids.[3] At the same time, measures were taken to define clearly the limits of each property, with the creation of a Topographic Commission in 1824, and the passing of several laws establishing the obligation on the part of the landowners to measure and delimit their property.

Also, by then the first law of *enfiteusis* had been passed at a provincial level, establishing that public lands were not to be sold, and making them available to tenants on twenty-year leases. The main purpose of the law was to prevent public lands from being transferred into private hands. The government was going through financial difficulties and needed the security of land for a loan that was being negotiated in London, and expected the lease-money to be received from the tenants to become a regular income for the Treasury.[4] Enfiteusis seemed to Rivadavia and his supporters an appropriate way of putting land into exploitation without giving it away for good. Yet whatever the purposes of those who fostered the system, it actually contributed to the concentration of land into the hands of the large cattle estancieros and speculators, who accumulated leaseholds and eventually bought the land at very convenient prices.[5] Although soon enough the pressure for land led to the introduction of new legislation between 1825 and 1837, over one million hectares of public land had been let to private tenants who paid—if at all—very low rents for their holdings.[6]

The Rosas administration saw the expansion of cattle raising to an extent that had no parallel in the previous decades (see Chapter 1). The insecurity of a fluctuating international market, specially for hides, favored the consolidation of cattle raising on an extensive scale, which required relatively low investments in technology and land, and yielded very high profits.[7] Investments had to be concentrated in cattle, and thus abundant, cheap, and relatively secure land was required.

This craving for land found a quick response from the provincial government, headed by the representative and leader of his class, Don Juan Manuel de Rosas. Combining the needs to make more land available for the estancieros, to solve fiscal problems, and to reward his followers, Rosas pursued a systematic policy of transferring public lands to private hands, by selling it in freehold or giving it away as gifts, pensions or other forms of reward. But this overriding supply of land found the market short of customers.[8] Prices went down, and the government had to authorize payment in *letras de Tesorería* and later on even in cattle. The 1839 rebellion against Rosas in the south of the province gave him the opportunity to give away some of the land the government had been unable to sell, to those who proved loyal to his cause.

It has been estimated that 8,500 *boletos de premio* were issued, although a large number of them were not used and others were transferred either to established estancieros or to speculators. Soldiers and clerks of the administration who were given rights for holdings under a square league in size most generally sold their part, thus favoring concentration of property into the hands of the few.[9] Almost 2.5 million hectares were given away by these different laws, but by 1852 only 572,535 hectares had been duly registered by the new owners. After 1840, the government passed no new legislation authorizing the massive transfer of land, and it is obvious that pressure from the private sector to obtain additional measures in this way had stopped. When the regime fell in 1852, the provincial government decided to review past donations and by a law of 1858, all land grants made between December of 1829 and February of 1852 were cancelled. There was an important exception, however. Those who had been rewarded on account of their participation in the wars against the Indians could keep their lands, and any of them who had not yet registered their respective claims to the land granted by law were given an additional period of time to do so and become legal owners. On account of this law almost 400,000 hectares were kept by private owners, while half that extension went back into the hands of the provincial government.[10]

The result of all these measures was the expansion of latifundia, and the concentration of land for the province as a whole. North of the Salado River we shall find that the situation was not homogeneous in all counties, as those that were closer to the capital or part of the *corredor porteño*—which connected Buenos Aires to the interior—had a more diversified economic structure, with agriculture and commercial activities being almost as important as cattle raising.[11] Those counties had been partially settled by the Spanish, and their land had been occupied from an early date, its concentration being therefore less likely than in the rest of the province. In the other counties, however, although the frontier had been kept at the Salado River for quite a long time, the structure of property and tenure shows a predominance of large-scale holdings, presumably entirely dedicated to cattle raising. The Cadastral Map of 1836 clearly shows this situation. Of the thirty counties north of the Salado River, I have selected sixteen[12] to carry out a detailed survey of landed property according to the Cadastral Maps of 1836, 1864 and 1890. The first map includes also holdings held in enfiteusis, so for this particular case I shall not only be speaking about private property but also about public land under the system of enfiteusis. Although the period we are interested in starts more than ten years after this map was made, I am using it in order to compare the structure of property that was predominant in the era of large cattle estancias with that which was to come about together with the diversification of production and the expansion of sheep breeding in the area.

Based on the cadastral information of 1836 I have prepared Lorenz curve I (Figure 3a) to show the distribution of land. As it was built on the basis of the data for different *partidos,* and some of the landowners had properties in more than one of the counties taken into account, I have constructed Lorenz curve IV (Figure 3b) in which I have added up holdings belonging to the same person.[13] In order to have a better idea of the concentration of land in the hands of certain social groups, I have also added up in this case land belonging to close relatives in different partidos. This information is probably far from complete, since I have included only those cases in which I could establish with certainty that the land belonged to the same individual or to close relatives (parents, brothers/sisters, husband/wife, sons/daughters). A further source of error may arise from the fact that 25.86% of the

Figure 3a

Structure of Property in Sixteen Counties
of the Province of Buenos Aires,
North of the Salado River,
1836, 1864, and 1890

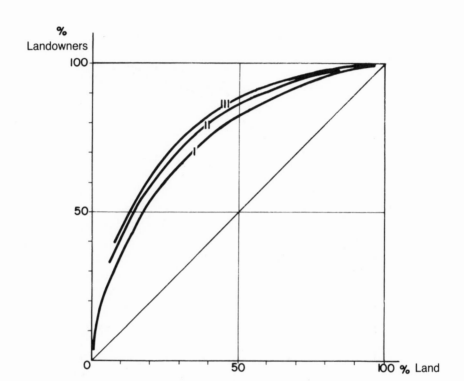

I 1836

II 1864

III 1890

Source: Sabato, op. cit., App. III, tables II, VI, VII.

Figure 3b

Structure of Property in Sixteen Counties
of the Province of Buenos Aires,
North of the Salado River,
*Considering Land Belonging to the Same Person
or to a Close Relative in One or More Counties
as One Holding,* 1836, 1864, and 1890.

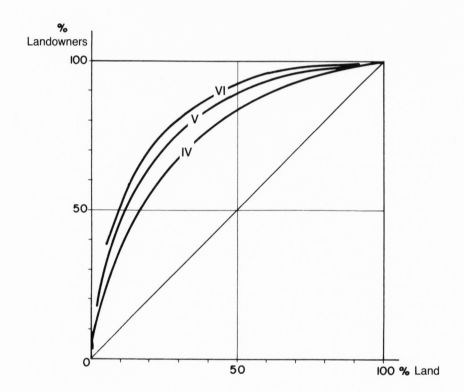

IV   1836

V   1864

VI   1890

Source: Sabato, op. cit., App. III, tables III, VIII, and IX.

Table 3
Holdings over and under 5,000 Ha in 16 Counties
of the Province of Buenos Aires

	Number of Holdings				Number of Hectares			
	Over 5,000 Ha		Under 5,000 Ha		Over 5,000 Ha		Under 5,000 Ha	
Year	No. of holdings	%	No. of holdings	%	No. of hectares	%	No. of hectares	%
1836	112	41.64	157	58.36	1,382,122	76.89	415,346	23.11
1864	122	14.04	747	85.96	1,228,884	51.03	1,179,375	48.97
1890	76	4.36	1,664	95.64	764,140	31.91	1,630,476	68.09

Source: Hilda Sabato, *Wool Production and Agrarian Structure in the Province of Buenos Aires, North of the Salado, 1840s–1880s* (Ph.D. dissertation, University of London, 1981), App. III, tables II, VI, VII.

land in the map appears to have no private owners or legal tenants, and therefore I have assumed that it is unoccupied public land, and have left it out of my calculations.

Cadastral information shows the following:

a. Large properties over 5,000 hectares—latifundia— were predominant, as they constituted 76.89% of the holdings (see Table 3). The mean of the universe is of 6,657 hectares, and the median lies within the interval of 3,750 to 5,000 hectares. Furthermore, only 4.85% of the land was held in holdings under 2,500 hectares.

b. Concentration of land was also a feature of this age: 19.3% of the landowners held 56.48% of the land, or 49 families had control of 1,015,255 hectares. Yet comparing our data with that presented by Carretero[14] for 1830 and 1851, it seems that for the area under consideration concentration of land was less than for the province as a whole, thus suggesting that south of the Salado River the presence of latifundia was even more striking.

c. North of the Salado River in those partidos which were closer to the capital and subject to earlier settlement, holdings under 5,000 hectares were predominant, while the highest proportion of latifundia is to be found among the counties of the south and west. Thus, in Baradero, Mercedes, San Antonio de Areco and San Vicente over 50% of the land is divided into holdings under 5,000 hectares, while in Lobos, Monte, Rojas, and Salto over 85% of the properties are over that size.

With this brief account in mind, we shall now go into the period we are specifically concerned with—i.e., when sheep breeding expanded throughout the province, and the agrarian structure as a whole was to be transformed.

1840–1890: Fifty Years of Change

As regards land, significant changes were to take place in the province of Buenos Aires, north of the Salado River, during this period. The transformation that was taking place in the productive structure had its parallel in the structure of property, and land (that basic means of agrarian production) was once more to be both subject and object of this transformation. Thus, we shall see how, between the 1840s and the 1880s, there was an effective appropriation of large tracts of land, a significant transference of property, and a relative expansion of medium- and large-size holdings (under 5,000 hectares). We shall also see how access to land that was once relatively easy, open, and cheap gradually became more and more limited and expensive, and how the larger landowning families, although fewer in number, ended up by concentrating relatively more land in their hands.

Although the Rosas administration had seen the expansion of cattle raising and a massive transfer of public lands into private hands, by the late 1840s Buenos Aires was still a sparsely occupied province, with many hectares of public land, and a part of its territory under Indian control. The existence of these lands, as well as the practice of leasing large extensions of them through systems like the enfiteusis had a decisive effect in the market, keeping prices relatively low and creating expectations as to the incorporation of new territories into productive use.

Public Land. We have said that Rosas' policy of alienating public land had resulted from a combination of political, economic and financial reasons. During the decades that followed his downfall, the same reasons would lead to very similar results. Successive laws were to bring about the transfer of lands from public to private hands, until by the 1880s the province had almost exhausted its reserves of this key resource.

After the fall of Rosas there was much discussion regarding the land policy to follow, and the matter was an issue in public debate. Whether or not to confiscate lands that had belonged to Rosas and his followers and how to proceed with property that had been acquired through boletos de premio were immediate questions the government had to answer. Yet beyond these practical matters, some very important policy issues arose, such as whether to keep the enfiteusis system or to replace it by other forms of lease, and whether to abolish tenancy altogether and have all land transferred into private property. Any discussion on land policy involved broader questions, and at the background of this whole debate lay the problem of what kind of future the ruling classes envisaged for the country. At this point, it is important to stress that although different groups within these classes did not necessarily agree on the answer to that crucial question, the actual policies which were followed in matters of land show a surprising continuity.

The trend among the different governments that came to power in the period under study was to favor private ownership of land, and even when the system of leaseholds was praised, it was done so because it was considered as the natural prelude to private property.[15] Not a few of the men who were to gravitate to the provincial political arena in this period held strong views about the need to favor the settlement of the population—especially immigrants—in the rural areas, to make land available for colonization schemes, and to discourage speculation. Men like Avellaneda, Sarmiento, Rufino Varela and C. Casares were inspired by the American model of frontier colonization. Yet the actual results of the official policy in matters of public land were to differ greatly from that ideal model. According to the existing studies on distribution of public land, speculation and latifundism were the corollary of these fifty years in the province of Buenos Aires.[16]

Several factors contributed to that result. From the point of view of the government—whatever the opinion of some of its spokesmen on the problem—land represented an important resource which could be sold or leased for fiscal purposes. Actually, the different land laws earmarked funds to be received by the provincial government from the operations they regulated.[17] Yet this explains only one aspect of the problem: the lack of a coherent set of measures devised according to the principles so many times mentioned in official documents and speeches. The urgent need for funds led to improvised and hurried legislation which sometimes did not prove rewarding even from the fiscal point of view.

But the results of this legislation were not confined to the formal content of these laws. During the first decade of the Rosas regime, the ascendant estancieros pressed for abundant and cheap land and were well rewarded by an administration very much identified with their interests. In the pastoral age, the rapid expansion of production upon a limited territory where private property had become the requisite for effective appropriation pushed prices up in the fertile counties which were also those better located in relation to the market (see below). In this sense, the situation of the province of Buenos Aires at the onset of the capitalist age was different from that found in other areas where land had been occupied for centuries and landowners were a well-established class when capitalist relations penetrated the rural areas. In the case of the latter, rent—no longer a relation of production—was inherited by the capitalist system as a relation of distribution.[18] In the province of Buenos Aires, land had been an abundant and cheap resource until the first decades of the nineteenth century, and it was only when the country started to export primary goods to the international market that land became crucial. The expansion of agrarian production for export purposes made land an indispensable, yet limited, means of production which, if privately owned, could allow for the realization and appropriation of rent. Thus, it was not the existence of a landowning class that made capitalist rent necessary, but the possibility of claiming such a rent that lay at the basis of the creation of a class of landowners.

Actual and expected increase in rent, and therefore in land prices, turned landownership into an attractive business. In the first place, it was a secure long-term investment which did not imply the immobilization of capital, as land could be mortgaged, and after the creation of the Banco Hipotecario it was increasingly so as this practice became widespread in Buenos Aires. Secondly, land could be the source of large profits if bought when it was still part of frontier territory and sold when the area was fully incorporated into production, speculating with the difference in price between those two moments.[19] Finally, and perhaps most importantly, landownership played a key role in the pattern of accumulation developed by estancieros in the second half of the nineteenth century.

Therefore, there were good reasons for pressing the government to sell cheap and for buying large tracts of public land when it was opened to private property. Those who belonged to the higher ranks of the River Plate society saw and made use of such opportunities, which in turn reinforced the reasons which made landownership an appealing proposition. Thus, if it is true that the initial distribution to the favored few gave the monopoly upon this increasingly required resource to those few, it is evident that a land market soon developed that could have challenged the original pattern of tenure. But this did not happen. Land changed hands and in many cases it was even subdivided, but it remained concentrated. And although there is a certain amount of truth in the argument that there were geographical and technical reasons for organizing cattle- or sheep-raising enterprises on a relatively large scale,[20] this cannot be extended to explain latifundism as it developed in the province of Buenos Aires. It is the pattern of accumulation which was at the basis of concentration and of its persistence beyond the initial stage of politically guided distribution. Favored by that initial distribution, pioneer estancieros organized their enterprises by counting on cheap and abundant land and devising a system of production that was based on an extensive use of that resource and a pattern of accumulation at the level of each enterprise that resulted from combining rent and profit in such a way that for every estanciero landownership became an integral part of his business (see chapter 4). This combination proved rewarding when prices of land rose in the early 1860s and again in the following decade, thus discouraging any fundamental changes in the system once the land market was in full operation. Therefore, there was no massive subdivision of plots even when demand pushed prices up high during the years of sheep expanion.

This is probably why projects like those of Sarmiento and Casares[21] to foster the development of agriculture and the division of property were doomed from the very beginning. In fifty years, the only achievement the province could show in this direction were the colonies of Baradero, Chivilcoy and Olavarría. And when the pattern of accumulation dominant during the pastoral boom started to show declining results and agriculture finally entered the province of Buenos Aires, it was also mainly carried out in land-extensive enterprises. As explained by Jorge F. Sábato, a new system of production was devised by the estancieros of Buenos Aires, but once more accumulation was based on the combination of both rent and profit, while landownership remained as an integral part of their business.[22]

By the 1880s most of the public land of the province had been given away, to the benefit of *latifundistas*—whether absentee landlords, capitalist estancieros, or (more frequently) both—and speculators, who concentrated large extensions of land in their hands. The legal instruments through which this was done took the form of a number of laws and regulations passed by both the national and the provincial administrations (see Appendix).

Although most of those measures related to the province as a whole, it is obvious that not all the areas were affected in the same way by legislation on public lands in the period under consideration. North of the Salado River there were not many leagues of public land left after the sales and gifts of the Rosas administration, and the laws of 1857 and 1867 probably completed their transference to private hands. The Cadastral Map of 1864 shows very little territory in this area still under state control. According to the lists published by Oddone,[23] of the 3,366,900 hectares sold as a consequence of the Law of 1836, 619,326 hectares belonged north of the Salado River. It is interesting to note that the average size of the holdings sold in these counties (5,385 hectares) was considerably smaller than the average for the province as a whole (14,327 hectares). In the lists for 1857, however, we find only 44 tenants in the counties chosen north of the Salado, and these mostly in the frontier partidos of Rojas, Pergamino and Salto. Although the average size of the holdings they rented was still smaller than the average for the whole province (7,987 hectares as against 9,900 hectares per head), they were larger than the 1836 average north of the Salado.[24] Those were still frontier lands, far from the capital, with a very deficient system of communications

and a scarce population, where agriculture was virtually unknown, and the estancieros required the cheap land in order to
carry on stock raising on an extensive scale.

The Prices of Land. The existence of large supplies of public land
naturally affected the market, and so did the government policies
towards that land and its prices. In most cases, laws regulating
leases or sales also at the same time established the respective
prices, or at least set the minimum value, as in the case of auctions
(see Appendix). Two different criteria prevailed as to the valuation
of public land—one assigned a uniform price to all lands within
the frontier and another price to all those lands beyond that line;
the other also took into account the location of those lands, especially their distance from the city of Buenos Aires. Both methods
were strongly criticized at the time because they made no dis-
tinction regarding the quality of the soil.[25] Another source of com-
plaint was the already mentioned fiscal preoccupation which lay
at the basis of these laws, making prices a function of the requirements of the Treasury.[26]

As regards private lands, their prices differed according to location and distance to market, quality of soil and natural advantages of the place, investments already made therein, and situation
in relation to the frontier. It is thus not possible to build a series
of single prices for land. Therefore I have tried to construct a table
containing the average and the range of prices to be found in the
30 counties under study, between the 1840s and the 1880s and
according to different sources. The rising trend of prices is striking. However, it should not surprise us, as it was the immediate
consequence of the incorporation of the province into the international market as a supplier of raw materials, and of the increasing importance of land as a source of rent.[27]

With the rapid expansion of production and trade in the pastoral
era, the pressure upon land grew steadily so that both actual and
expected rises in land rent pushed prices upwards in the long
run. However, this trend had frequent oscillations. The sharp
fluctuations in the supply of land resulting from an ever-moving
frontier and from the large extensions still in public hands, plus
the ups and downs of a demand governed by such varied factors

as speculation and the international market for wool and hides, all tended to determine day-to-day changes in the prices of land.

Thus, with the expansion of sheep raising and wool exports in the 1850s, prices of land experienced a steady increase, which became a sharp one in the prelude to the crisis of the mid-sixties. In Chapter 1 we mentioned the main aspects of that crisis in the area under study. As regards land, overstocking was the main problem, arising from the expansion in the size of the flocks and the relative stagnation of the frontier in the areas of soft grasses suitable for sheep raising during the previous decade. A severe drought was a further setback for cattle and sheep owners, many of whom had to take their herds to the neighboring provinces of Santa Fe and Cordoba, in order to avoid catastrophe.[28] Contemporary sources considered the Law of November 1864 regarding public lands an aggravating factor to the crisis, claiming that prices established by that law were exceedingly high.[29]

The slump that followed this period of land hunger is shown clearly in Table 4, and contemporary observers claimed the value of property came down by 50%.[30] Yet the quick recovery in the following years probably helped to compensate for this decrease, and prices continued to go up until the crisis of 1873–75. The effect of this crisis on the price of land can be seen in Table 4, in the figures for the period 1875–79, which are lower than those for the previous five years. Nevertheless, once more recuperation followed suit, and up went the prices for land.

North of the Salado River it is evident that by the 1880s land in private hands had developed into an important source of rent, as production and exports expanded. Provided with soft grasses and good watering, the area was best suited for sheep raising, although some of the counties immediately south of the river shared that privileged situation. A still deficient system of communications and transportation sharpened the differences between lands located—in absolute and relative terms—closer to Buenos Aires and the railway network, and those located in the far south and west. Therefore, the price of land was ever increasing in that part of the province, where we also find an intensification of its productive use.

In Chapter 1 we have seen how cattle and sheep stocks grew in the area and period under consideration. This resulted both in

Table 4
Prices of Land[a] in the Province of Buenos Aires,
North of the Salado River (in $oro per Ha)

Years	Average prices	Range of prices
1843–49	.68	0.33 to 1.00
1850–54	1.11	0.65 to 1.48
1855–59	5.76	0.67 to 8.83
1860–64	12.04	2.15 to 17.45
1865–69	7.46	4.44 to 19.77
1870–74	15.11	6.92 to 32.16
1875–79	13.69	4.94 to 28.30
1880–84	19.20	7.00 to 34.50

[a]Does not include prices of public land.

Sources:

Archivo General de la Nación, Sala III, *Libros de Contribución Directa*, Campaña, 1865 and 1867.

Archivo General de la Nación, *Protocolos Notariales*, a sample of 90 documents for the years 1850–1880.

Provincia de Buenos Aires, *Censo General de la Provincia de Buenos Aires*, 1881 (Buenos Aires, 1883).

H. Gibson, *The History and Present State of the Sheep-Breeding Industry in the Argentine Republic* (Buenos Aires, 1893), 96–103.

T. Hutchinson, *Buenos Aires y otras provincias argentinas* (Buenos Aires, 1945), 328.

W. Latham, *The States of the River Plate* (London, 1868), 215.

MacCann, *Two Thousand Miles Ride through the Argentine Provinces* (2 vols., London, 1853), vol. 1, 8, 20, 86, 143, 273.

Parliamentary Papers, Commercial Reports (b) Embassy and Legation, vol. LXIX, 1867, 329, and 1881, vol. LXXXIX, 163.

B. Vicuña Mackenna, *La Argentina en el año 1855* (Buenos Aires, 1936), 410.

C. F. Woodgate, *Sheep and Cattle Farming in Buenos Aires* (London, 1876), 20–22.

the incorporation of most land available into productive use and in the more intensive utilization of that land. Thus we have seen that stocks rose not only in absolute terms but also in the number of heads per hectare. Yet this was done at the expense of the quality of the soil because, as we shall see in Chapter 4, little was done to preserve or improve the natural condition of that soil. The age of fodder crops was not far off; it would reach this area in the late 1880s. Overstocking became critical in certain periods, although the possibilities of expanding beyond the established frontier sooner or later tended to alleviate the situation in the more densely stocked areas.

The Structure of Property. Although there were different ways of incorporating land into productive use, private property became more and more the precondition for its appropriation. We have seen how in the 1820s and 1830s, different public agencies were put into operation to oversee private property, and numerous laws were passed in those and the following decades to ensure the transfer of public land into private hands. Thus, little room was left for occupation without a legal title.

The land tenure system originated in the initial distribution of public lands to the members of the propertied classes of Buenos Aires and was later modified by private transactions in the market that was shaping up in this period. In the following pages we shall see how this system evolved between 1836 and 1890, and what were its main features during the pastoral era.

In order to analyze the structure of property in the period and area under consideration, I have worked on the Cadastral Maps of 1864 and 1890, as mentioned when describing the situation for 1836.[31] To portray the distribution of land, I have built Lorenz curves II and III and Lorenz curves V and VI, where holdings belonging to the same person or to close relatives in one or even several counties, have been added up and considered as a single property (Figure 3a and 3b). Although this has been done only in those cases in which the relationship is certain, and therefore many cases might have been left out, the operation will allow us to see the degree of concentration of land in the hands of family groups.

As the limits of the different counties changed more than once

during the period under study, in order to make comparisons possible I have taken the limits of 1890 as valid for the three years in which we have cadastral information. Therefore I shall always be speaking of the counties in their 1890 spatial delimitation.
This cadastral information suggests the following:[32]

a. Between 1836 and 1864, and again between 1864 and 1890 the number of holdings increased quite significantly, from 270 for an occupied territory of 1,797,468 hectares to 869 for 2,408,259 hectares and again to 1,740 for 2,394,616 hectares, bringing down the average size of the holdings. Thus, if for 1836 the mean of the universe was 6,657 hectares, and the median lay in the interval of 3,751 to 5,000 hectares, for 1864 both had declined to 2,771 hectares and 1,001 to 1,750 hectares respectively, further to go down for 1890 to reach 1,376 hectares for the mean, the median being within the interval of 501 to 1,000 hectares.

b. Latifundia, relatively predominant in 1836, tended to decline in the following decades, while the opposite was true for holdings under 5,000 hectares (Table 5).

c. Subdivision of landed property was a significant feature during the period under consideration, as is clearly shown by Table 3 and by observing the cadastral maps. We shall see below that this process took place through the transference of land both by inheritance and by purchase.

d. The structure of property was not even throughout the sixteen counties taken into account in this study, although the trend toward subdivision of property and the predominance of smaller holdings over larger ones was a general one for all the area under consideration. However, Monte, Rojas, and Salto, together with Arrecifes and Chascomús, continued to be in 1864 the counties with the largest proportion (60 to 70%) of the holdings over 5,000 hectares, while by 1890 they accounted for 47 to 57% only in Chascomús, Salto and Monte. As for property under 5,000 hectares, Mercedes, Navarro, San Vicente and Suipacha had over 70% of their holdings under that size by 1864, the percentage rising to over 80% in those counties, as well as in Baradero, Arrecifes and Ranchos by 1890. These differences between counties had to do with the structure of production in each place, and with the nature of the soil, the distance to Buenos Aires and other markets, the date and characteristics of settlement, and other such factors.

e. As for the distribution of property, it is interesting to compare the Lorenz curves for the years 1836, 1864, and 1890 (Figure 3a and 3b), as they show an increasing separation from the line of equidistribution. This suggests that land was more evenly distributed during the first decades herein considered, and that polarization increased with time. Therefore, although the number of large holdings and the total amount of land under the control of latifundistas may have diminished significantly, inequalities were enlarged, and fewer men had the ownership of relatively more land. This feature appears once more if we consider Lorenz curves IV, V, and VI, where all the land belonging to a certain family has been added up as if it constituted a single holding.[33]

Cattle raising carried out on an extensive scale required large properties, and therefore in 1836 holdings were more evenly extensive. But with the advent of sheep raising (and later on with agriculture) it was possible to start production with holdings even under 1,000 hectares and many enterprises were set up in such conditions. Large estancias not only persisted from the previous age, however, but they developed as a result of a system of production which still rewarded an extensive use of land. And though sheep growing could and did prosper on holdings considerably smaller than those predominant in the cattle age, higher returns were obtained in larger units. The expansion of small and medium-sized holdings together with a persistence of large estancias resulted in a pattern of land distribution more uneven than the one for 1836, and this showed increased polarization within the propertied classes.

With this description of the structure of property at three specific moments, we have something like snapshots of the reality we are trying to analyze, but as yet we cannot establish any connection between the different structures, except from a comparative point of view. In order to study the property of land in its evolution throughout the period under consideration, I have explored the origin and development of the holdings, tracing how much of the land in private hands in 1864 and 1890 was kept by the same families—probably inherited—from 1836 and 1864 respectively. As I have done the calculations on the basis of the Cadastral Maps, and have included as inherited cases only those in which I could be certain that the land had remained in the hands of the same family, there is probably an underestimation of the figures for inherited lands, as sometimes it was not possible to trace the origin of the ownership of certain holdings.[34] For the sixteen counties as a whole, the percentage of land which was held in 1864 by the same families as in 1836 appears to be only 17.61%, suggesting that for the first decades of our study and for this area north of the Salado, transference of land was a common practice and access to property was not limited to those already in the business. This proportion increased in the following stage, but even so 43.22% of the land seems to have changed hands between 1864 and 1890. Even allowing for underestimation in the figures for inherited land, they suggest the existence of a relatively open market for that resource, which tended to become more rigid as the years passed.

In order to study in more detail the source of landed property in the counties of our sample, we have identified the families who, according to the consulted maps, owned over 10,000 hectares within the area in any of the three years considered, and have tried to trace the origin of their property. By analyzing the cases of these few families we are, however, taking into account a very large proportion of the lands under consideration, as well as a group of landowners who had a relevant role to play in the period and area under study. Their number and relative importance are stated in Table 5, while the origin of their property is shown in Table 6.

From these tables and from the list of families we get the following information:

a. By 1836, most of the land was either held in enfiteusis or had been bought from the state by former enfiteutas. Nevertheless, renting public land through that system did not necessarily mean its purchase, and we see that in 11 out of 34 cases shown in Table 6, the right to the land was eventually lost or transferred back to the state or to others.

b. By 1864, a large number of properties still originated in old

Table 5

Families Who Owned over 10,000 Ha in 16 Counties
of the Province of Buenos Aires, North of the Salado River,
in 1836, 1864, and 1890

Year	Number of families	Percentage of total	Number of hectares	Percentage
1836	49	19.30	1,015,255	56.48
1864	51	7.10	1,004,643	41.73
1890	44	3.59	875,157	36.55

Source: Hilda Sabato, op. cit., App. III, tables III, VIII, IX.

Table 6

Origin of Properties over 10,000 Ha Identified in Table 5[a]

Year	Enfiteusis and law of 1857[b]	Gifts and donations	Purchase[c]	Inheritance[d]	Unknown
1836	34[e]	3	4	—	8
1864	17	1	10	10[f]	13
1890	—	—	9	29[g]	6

[a] Most properties were formed by land acquired through different methods, but in this table I have tried to identify the initial source of property, although the final holdings probably resulted from a combination of sources.

[b] Includes families who had access to land in the area under consideration initially as tenants of public lands.

[c] Excludes purchase from the state, which is included in the first category.

[d] Includes only inheritance within the universe of owners of over 10,000 Ha.

[e] Of these 34 enfiteutas of 1836, later on 11 transferred or lost their rights to the land, 15 bought the land from the state, and 8 we have no information about.

[f] Of these, 7 were among the enfiteutas identified in 1836.

[g] Of these, 17 were among the enfiteutas identified in 1836 and 1864.

Sources:

Cadastral maps of 1836, 1864, and 1890.

Casa Pardo, *Nuestras estancias* (Buenos Aires, 1968).

D. Hernando, *Casa y familia* (Ph.D. diss., University of California, Los Angeles, 1973).

M. G. and E. T. Mulhall, *Handbook of the River Plate Republics* (Buenos Aires and London, editions of 1868, 1875, and 1885).

R. Newton, *Historia de la Sociedad Rural Argentina* (Buenos Aires, 1966).

J. Oddone, op. cit.

A. Carretero, "Contribución al conocimiento de la propiedad rural en la provincia de Buenos Aires para 1830," *Boletín del Instituto de Historia Argentina Dr. E. Ravignani*, 13, No. 22/23 (2nd series) (1970), 273–292, and *La propiedad . . .* , op. cit., 60–160.

enfiteusis rights (23 in all, as only one holding sprung from the 1857 Tenancy Law).

c. By 1864 only ten of the families identified in 1836 still held over 10,000 hectares in this area, while by 1890 the number had gone down to eight. The period of 1864 to 1890, however, shows more continuity, with 29 of the families of the former year still among the latifundistas of the latter.

d. Purchasing appears as a viable way of access to land, both for 1864 and for 1890. Yet we find that out of nine of the holdings over 10,000 hectares purchased before 1890, seven were at least partly acquired before 1870. We also find that 41 out of the 44 holdings identified in 1890 were at least partly in the hands of their respective owners before 1870. This suggests that the latifundistas of the 1890s were no newcomers to the rural areas, and that their interest in land had started at least twenty years earlier. At the same time, very few of these families had had anything to do with land before the 1830s, suggesting that most fortunes in land were the product of those forty years of expansion.

e. As to the main business interests of each of these families, most of the landowners and enfiteutas identified in 1836 were, or had been, merchants, officers of the administration (both colonial and post-independence) and high-ranking military officers,[35] while both in 1864 and 1890 their main branch of business was generally sheep and cattle raising, and not a few of them were also involved in financial and commercial activities connected to the export economy. Most of these large landowners were not absentee landlords who kept their holdings as a source of rent, but were rural entrepreneurs, who organized sheep or cattle estancias so as to receive both rent and profit, thus becoming capitalists and landlords. Thus, in 1890 out of 44 owners of over 10,000 hectares north of the Salado, at least 29 were sheep estancieros. It is important to point out that most of these families also owned land in other areas of the province and of the country.

Some Final Remarks

In the transformation that was to take place between the 1840s and the 1880s in the Argentine society the province of Buenos

Aires had a crucial role to play as supplier of the main products that were to be exported to the international market. We have already seen how production expanded in those fifty years in the area north of the Salado River, and how sheep raising gradually became the main activity in this area as exports of wool increased, and the pastoral industry developed into a profitable business.

As one of the main aspects of any process of agrarian change, land was not foreign to these transformations. By the first decades of the century, large areas of land in the province of Buenos Aires had been the realm of Indians, wild cattle and birds. North of the Salado some areas had been nominally appropriated by men from Buenos Aires, but very little had been actually settled. Land belonged mostly to the state, but its occupation was open to almost anyone who wished to do so.

By mid-century, however, the situation was changing rapidly. The development of cattle raising, and later on of sheep breeding, led to a conscious effort on the part of the state to ensure the effective appropriation and incorporation to productive use of the whole territory of the province. Consolidation of private property, transfer of public land into private hands, and expansion of the frontier through the extermination of the Indians, were the means used to achieve that end.

Land therefore became private, and gradually this resource of easy access, cheap and abundant in the 1840s and early 1850s, by 1890 had developed into an expensive and scarce means of production. North of the Salado, all land had been transferred to private hands by the 1890s. The expansion of sheep raising had led to the effective occupation of most of the territory, and would soon foster the subdivision of property, while prices of land swelled. Latifundia, predominant in the 1830s and 1840s, had diminished significantly by the 1880s, but polarization nevertheless increased. The laws regulating inheritance also favored subdivision, and although they could be evaded by the formation of joint stock companies, this hardly affected property in the period and area under consideration.

So, by 1890, the desert had been transformed. No more land belonged to the state, no more territory north of the Salado River remained empty. In sixteen counties of that area, 1,740 landowners had replaced the initial 269, and the average holding was much smaller than its predecessor had been. The few who owned over 10,000 hectares still had control of a large part of the territory, but most of them did not belong to the same families of the 1830s. Land had changed hands in this period, yet distribution remained as uneven as ever. Subdivision had been the product of the last fifty years, but so was accumulation. *Why* and *how* are questions for future chapters.

Notes

1. Andrés Carretero, *La propiedad de la tierra en la época de Rosas* (Buenos Aires, 1972), 13.

2. Tulio Halperin Donghi, "La expansión ganadera en la campaña de Buenos Aires," in T. Di Tella and T. Halperin Donghi (eds.), *Los fragmentos del poder* (Buenos Aires, 1969), 56.

3. Miguel A. Cárcano, *Evolución histórica del régimen de la tierra pública 1810–1916* (Buenos Aires, 1972), chap. 3.

4. Ibid., p. 40.

5. See, for example, Cárcano, op. cit., and Jacinto Oddone, *La burguesía terrateniente argentina* (Buenos Aires, 1967).

6. Tulio Halperin Donghi, "La expansión de la frontera de Buenos Aires," in Alvaro Jara (ed.), *Tierras Nuevas* (México City, 1969), 80.

7. Ibid., 83.

8. Myron Burguin, *Aspectos económicos del federalismo argentino* (Buenos Aires, 1969), 321–322.

9. Carretero, op. cit., 25–30.

10. M. E. Infesta de Guerci and M. E. Valencia de Placente, "Tierras, premios y donaciones," paper read at the VII Jornadas de Historia Económica, Rosario, Argentina, Oct. 1985 (mimeo); and M. E. Valencia de Placente, *La política de tierras públicas después de Caseros (1852–1871)— Provincia de Buenos Aires* (Ph.D. dissertation, University of La Plata, 1983).

11. Halperin Donghi, "La expansión ganadera," 42.

12. The selection of counties for the sample took into account both their location—to ensure an even regional representation north of the Salado River—and the quality of the Cadastral Map for 1890, where the measurements were done. The limits for the counties are always those of 1890. The selected counties are the following:

North: Arrecifes, Baradero, Pergamino, Rojas, San Antonio de Areco, San Pedro.

Center: Carmen de Areco, Lobos, Mercedes, Navarro, Salto, Suipacha.

64 ◊ Chapter 2

South: Chascomús, Monte, Ranchos, San Vicente.

13. Cadastral information was used to build several tables which show the structure of landed property in the selected counties. These tables are included in Hilda Sabato, *Wool Production and Agrarian Structure in the Province of Buenos Aires, North of the Salado, 1840s–1880s* (Ph.D. dissertation, University of London, 1981), chap. 2 and App. III.

14. Carretero, op. cit., 14.

15. See, for example, Bartolomé Mitre, *Arengas* (Buenos Aires, 1889), in particular pp. 151–152, "Speech of Sept. 16th, 1857."

16. See, for example, M. Bejarano, "Inmigración y estructuras tradicionales en Buenos Aires, 1854–1930," in T. Di Tella and T. Halperin Donghi (eds.), *Los fragmentos del poder;* Cárcano, op. cit.; E. Coni, *La verdad sobre la enfiteusis de Rivadavia* (Buenos Aires, 1927); H. S. Ferns, *Britain and Argentina in the Nineteenth Century* (Oxford, 1960); Oddone, op. cit.

17. Bejarano, op. cit., 86; Cárcano, op. cit., chaps. 10, 11, 19; *Le Courier de la Plata*, 6 May 1867. See also Appendix.

18. See Pierre P. Rey, *Les alliances de classes* (Paris, 1975).

19. Speculation in this period took two main forms—the purchase of public land at convenient prices to be sold later at profit, and the subdivision and subletting of holdings. This latter practice had already been denounced by Pastor Obligado in 1855, when he pressed the legislature to rule on matters of public land, and was mentioned in different laws that established priority of purchase for the occupants of a plot (Mensaje del Poder Ejecutivo a la Segunda Legislatura Constitucional del Estado de Buenos Aires [Buenos Aires, 1855], 15).

20. For this line of argument see, for example, Roberto Cortés Conde's book, *El progreso argentino, 1880–1914* (Buenos Aires, 1979), esp. chap. 2.

21. I am referring here to Sarmiento's project of creation of agricultural centers alongside the Ferrocarril Oeste (early 1860s), and to Casares' Law of 1876 to promote the settlement of population in rural areas.

22. Jorge F. Sábato, *La clase dominante en la Argentina moderna: Formación y características* (Buenos Aires, 1988).

23. Oddone, op. cit., 101–109.

24. Ibid., 126–136, and Valencia de Placente, op. cit.

25. *Le Courier de la Plata*, 17 June 1866 and 22 June 1866.

26. Ibid., and Olivera, op. cit.

27. For different interpretations of the role of ground rent in the determination of prices and, in general, of the regulation of land prices, see, among others, S. Amin and K. Vergopoulos, *La cuestión campesina* (Mexico, 1975); A. Cutler, "The Concept of Ground Rent and Capitalism

in Agriculture," *Critique of Anthropology*, 4–5 (1975); G. Elichman, *La renta del suelo y el desarrollo agrario argentino* (Mexico City, 1977); M. Gutelman, *Structures et réformes agraires: Instruments pour l'analyse* (Paris, 1974); K. Kautsky, *La cuestión agraria* (Mexico City, 1974); K. Marx, *El Capital*, III (2nd ed., 5th rpt., Mexico City, 1972); P. P. Rey, *Les alliances de classes* (Paris, 1973); D. Ricardo, *Principles of Political Economy* (London, 1971); A. Scott, "Land and Land Rent," *Progress in Geography*, 9 (1976); K. Tribe, "Economic Property and the Theorization of Ground Rent," *Economy and Society*, 6, No. 1 (Feb. 1977), 69–88.

28. José C. Chiaramonte, *Nacionalismo y liberalismo económicos en Argentina (1860–1880)* (Buenos Aires, 1971), chap. 2, and Olivera, op. cit.

29. Olivera, op. cit., and *Le Courier de la Plata*, 17 June 1866 and 22 June 1866.

30. Chiaramonte, op. cit., 66 (quoting Olivera).

31. Buenos Aires, Provincia de, *Registro gráfico de las propiedades rurales de la Provincia de Buenos Aires*, construido por el Departamento Topográfico y publicado con la autorización del Superior Gobierno de la Provincia, 1864; and *Registro gráfico de 1890* (Buenos Aires, 1890). The measurement of the size of properties in the sixteen counties chosen among the thirty of our study, was done on the maps for 1890, which are drawn on a sufficiently large scale to allow for calculations to be done with a planimeter, with an expected error not exceeding 10%. For each partido, I have measured all holdings indicated on the respective map, and identified the owners. Calculations for 1864 were done on the basis of the 1890 maps, as the scale of the 1864 maps is not large enough to allow for direct measurement. Therefore, the results obtained for 1864 are probably not so reliable as those for 1890, although calculations have been done very carefully in each case. Comparisons with the 1836 map are possible, but we should always bear in mind that the estimates on that map may be far from exact, and that it was built to include land held not only in private property, but also in enfiteusis.

32. Tables showing this information are included in Hilda Sabato, op. cit., chap. 2 and App. III.

33. If we consider all the land held in properties of over 10,000 Ha, in the first case we have that for 1836: 18.58% of the owners have 52.84% of the land. For 1864 and 1890 the figures are 4.26% to 26.54% and 1.37% to 17.21%. In the second case (Lorenz curves IV, V, and VI) for 1836, 19.30% of the owners control 56.48% of the land, while for 1864, 7.10% have 41.73% of the land, and for 1890 the figures are 3.59% and 36.55% respectively.

34. For these calculations, see Hilda Sabato, op. cit., App. III, Table X.

35. This drive of the urban propertied classes of Buenos Aires towards the rural areas of the province has been described by, among others, Tulio Halperin Donghi in several of his works, particularly *Revolución y guerra: Formación de una élite dirigente en la argentina criolla* (Buenos Aires, 1972). For the case study of several families who followed that pattern, see Diana Hernando, *Casa y Familia: Spatial Biographies in Nineteenth Century Buenos Aires* (Ph.D. dissertation, University of California, Los Angeles, 1973).

3

Labor

Nada tan triste como la vida que esa miseria representa."
—Emilio Daireaux, *Vida y costumbres en el Plata*, 1888
(referring to the shepherd's life).

Sheep and land became the principal means of production in an agrarian structure dominated by the pastoral industry. In the process of production they were brought together by capital, which also employed the labor force required to furnish the necessary work that would transform raw materials into a finished product, whether it be wool, tallow, mutton, or skins. Before analyzing that process, which will be the subject of the next two chapters, it is my purpose to describe the main characteristics of the labor force that took part in sheep raising and wool production.

Although many of the problems raised in this chapter will be better understood after reading those concerning the units of production and the relations that developed therein, I found it necessary to make a separate analysis of the origins, characteristics, and process of reproduction of the labor force. A brief introduction will give a general view of the situation before the pastoral period, followed by a description of the changes observed in the pattern of labor demand since mid-century. I will then turn to the problems posed by the supply of labor, briefly referring to the structure of population in the period under study and analyzing the process of expansion of a wage labor supply to meet the needs of increasing demand. Finally, I will concentrate upon the specific problems of the labor force employed in sheep estancias and farms.

The Postrevolutionary Period

We have already mentioned that cattle raising on an extensive scale was the only significant productive enterprise carried out in the province of Buenos Aires during the first half of the nineteenth century. Abundance of land, and scarcity of capital and labor force led to the organization of large cattle estancias, which provided meat for the *saladeros* and hides, the staple exports of the province. Its territory was only sparsely populated as these estancias required relatively few men to work over large extensions of land. Small and scattered provincial towns provided for the local commercial needs, and sometimes were links in interprovincial or riverine routes. Thus, San Nicolás and Zárate, as well as Luján and Pilar, were relatively important since colonial times. The estancias carried out what little agriculture there was, generally for self-consumption, along with *quintas* and *chacras* established in the periphery of small towns. Nevertheless, as we have seen, this production was not enough to cover internal demand, and basic products, such as flour, had to be imported.

The picture, however, was far from static. Expansion of production required the occupation of new territories if its scale was to be kept extensive. Therefore, there was an appropriation of new lands, as the frontier moved forward, and cattle and men started to flow into these new areas. Halperin has portrayed this drive to the south, pointing out a reorientation of the population toward the frontier, and the massive incorporation of immigrants from the interior provinces.[1] Population grew rapidly in these new territories, but more slowly in the old established partidos. For the whole province, excluding the capital, we find 177,060 inhabitants in 1854. This is more than double the number of 1822, when they were estimated at 74,600, thus indicating a rate of growth of 2.74% per year.[2]

According to the Census of 1854, the estancias of the province were in the hands of a total of 9,856 owners and tenants, while the number of laborers (*peones*) was 20,313.[3] What were the main characteristics of this labor force? Although there is a large number of works on the first half of the nineteenth century, very few of the historians have gone into the problem of the organization of the estancias in the province of Buenos Aires in that period, or into the characteristics of the labor force used therein. The gaucho has been the subject of different kinds of analysis, ranging from the more anecdotal descriptions and literary insights of the contemporary observers to more recent attempts to study the social context in which the gaucho emerged. Nevertheless, no complete historical analysis of the role of the gaucho in the productive structure of Buenos Aires is to be found, nor of his transformation throughout the different periods in the rural history of the province.[4]

Those who have studied cattle raising and agrarian production have concentrated on its more general aspects, and perhaps only T. Halperin Donghi, E. Astesano, and more recently John Lynch[5] have explored the estancia regime of post-revolutionary Buenos Aires. Halperin has stressed the important role the estancia came to play in that period, which led to continuous attempts to tighten its internal organization, as well as to control effectively the labor force employed. In a situation characterized by an acute shortage of labor, the constant need to recruit men for the army—indispensable to secure new lands but also to take part in internal and international conflicts—had to be matched with requirements of hands by the estancias, and the legislation on *vagos y malentretenidos* seems to have favored both. Thus it was a means of recruiting the marginal and of putting pressure on the rest, who in the end preferred estancia work to the front.

Wage labor was predominant in cattle estancias. Generally, a *mayordomo* was left in charge of the organization of the work within the estancia, and he had *capataces* to control and direct the jobs being carried out by herdsmen and peones. Day-to-day, routine work consisted of taking care of the herds, and driving them to the *rodeo*, where they assembled every morning. Periodical operations such as cattle branding and horse breaking required extra hands hired for the occasion. Permanent laborers generally lived within the estancias, and were paid wages—part in cash and part in kind, mainly food and lodging. Temporary hands usually enjoyed relatively higher wages, as the scarcity of labor gave them a certain bargaining power vis-à-vis the estancieros.[6]

The cattle estancia would by no means disappear in the second half of the century, although its importance decreased north of the Salado, as sheep raising became predominant. In 1867, for example, Latham would describe a cattle estancia which undoubtedly had very few, if any, differences from its predecessor of the Rosas era:

The distribution of a cattle estancia "plant" is similar to that of a sheep farm: the estancia house, with horse corrales and cattle corrales, and puestos in different parts of the ground for the herdsmen. Each puestero has his herd (rodeo) of cattle and a tract of land appointed to him.

There is a "capataz" (overlooker) to a certain number of puestos, and a mayordomo, or manager, over all: there are also immense troops of wild mares and horses. A "rodeo" or herd, is various in size, consisting of a few hundreds or a few thousands semi-wild cattle.

In these estancias we see the true type of the gaucho of the Pampas—a type now-a-days rarely found in the sheep districts.[7]

Labor in Sheep Estancias

Large estancias, a few men, and semiwild cattle was the simplified equation of pre-1850 Buenos Aires province. Dealing with cattle and horses was the sole occupation of the countryside, a business requiring dexterity and skill, but only in a very limited number of jobs. Nevertheless, the shortage of labor was a constant feature of this period, and coercion in different disguises was exerted on the existing hands to ensure their employment and discipline. By mid-century, as we have seen, a shift in production began to take place. Gradually, new estancias were set up entirely dedicated to sheep raising, and many an old cattle establishment also gave way to that new and promising activity. This transformation brought about important changes in the pattern of labor demand.

In the first place, labor demand swelled. As it was organized in Buenos Aires, sheep raising required more hands per hectare than did cattle, but it was the rapid growth of the flocks that contributed most to the expansion of demand. Even though the productivity of labor increased with time, particularly after the introduction of wire fencing in the 1870s, our estimates are that labor demand rose at least ten times in thirty years (see Table 7).

Secondly, sheep raising required the exercise of skills formerly unknown to the native workers. In this new period, many of the workers to be found in an estancia would appear under the same names as before, yet their roles had greatly varied. From the mayordomo to the peon, new tasks were assigned and had to be

Table 7
Permanent Labor and Temporary Labor for the Shearing Season
Required by the Pastoral Sector in the Province of Buenos Aires,
Estimates for 1850–1885[a]

Year	Permanent workers	Temporary workers— shearing season
1850	3,000	2,000
1855	4,700	3,100
1860	6,700	4,600
1865	22,000	15,000
1870	27,300	18,000
1875	33,300	22,000
1880	36,000	24,000
1885	47,000	31,000

[a]These figures have been estimated on the basis of information regarding the number of sheep shorn in the provincial herds, calculated on the basis of data on wool exports and average clip per head; the number of all permanent workers required by estancias and farms (whether wage laborers, sharecroppers, or farmers) estimated at one man per 1,500 head; and the extra hands required during the shearing season (estimating the length of the season as 70 days and the number of sheep shorn per man per day as 40, and calculating that the operation required 5 men dedicated to complementary tasks for every 20 shearers). This table does not refer to the number of workers actually employed by the pastoral sector but to the number of hands *required* by it for different years in the period under study, and the figures represent only a very rough estimate of the actual demand of labor, as they do not take into account variations observed in the productivity of labor throughout the period.

learned when the organization of the establishment was adapted to the new type of productive enterprise. In the higher echelons of the internal order, those responsible for the management of sheep estancias had to attend to the problems posed by cross-breeding in order to obtain better wools, to supervise the curing and bathing of the sheep, to organize the shearing process, and, in short, to deal with an entirely different set of problems than those who were in charge of cattle establishments. Lower in the scale, each puestero had to take care of 1,500 to 2,000 head. According to Gibson:

> His duties are to tend his flock day and night; to keep it from mixing with other flocks which run on the same estate when there is no divisory fence between the several runs; to keep it free from scab and other contagious diseases; to keep dogs off, and see that no sheep wander astray; in short, to generally shepherd his charges. All this he does upon horseback.[8]

So, although the type of herdsmen and shepherds was similar in nature, the actual tasks performed by them differed greatly and required different skills. The work of the peon would change as well, as jobs like the curing of the sheep or shearing were unknown in the cattle age.

In the third place, the internal organization of the estancia became tighter than in previous decades. The annual schedule of activities had to be met with precision, and managers had to exert complete control over all tasks performed in the puestos and the head station. Moreover, it was imperative that shepherds stay at their working places, as there could be great harm for the establishment if they happened to leave the place without a substitute.

Finally, sheep raising had seasonal peaks of activity which deeply affected labor demand. Therefore, the pastoral sector not only required increasing numbers of permanent workers but it also engaged annually, during the shearing season, a large number of extra hands who were employed for the occasion.

Besides engaging permanent and seasonal workers, estancias also employed casual laborers for specific jobs, such as fencing, ditching, repairing buildings, and so on, as well as for more routine though irregular tasks, such as horse breaking, earmarking, and castration.

How did the supply of labor meet this pattern of demand? In order to analyze this problem, I shall first include a brief description of the structure of population during the period under study, and then refer to the main trends in the supply of labor.

The Structure of Population

It is a very large territory, and yet a small population has been and still is a striking characteristic of Argentina. The province of Buenos Aires, though favored by excellent geographical and natural conditions, enjoying political and economic supremacy throughout long periods in its history, and attracting large numbers of immigrants from all over the world, has shared this feature, even if to a lesser degree than other parts of the country.

At the turn of the eighteenth century, Azara estimated that only 32,168 people inhabited the province, outside the capital, and this indeed is a very small number, even allowing for the fact that its territory was four to five times smaller than the present one. By 1895 that population had risen almost thirty times to somewhat over 920,000, and yet it was still scattered and dispersed. In order to portray how this population grew and changed throughout the century, particularly in the area north of the Salado River, I have used the National Censuses of 1869 and 1895, the Provincial Censuses of 1854 and 1881,[9] and other statistical sources. By analyzing this information we observe the following:

(a) For the province as a whole, the population in absolute numbers increased steadily throughout the century (Table 8), but its rate of growth started to decline in the 1880s. For the counties under consideration and for the second half of the century, however, the rate of growth oscillated, reaching its peak in the interval 1869–1881, coinciding with the last period of sheep expansion. After a decline in the 1880s, the rate of growth went up again in 1890–1895, when immigrants were pouring in large numbers into the province as it shifted to agriculture and cattle production (see Maps 9, 10, 11).

As new lands were appropriated and put into productive use, the population tended to expand into the new territories, and therefore we find that counties in frontier areas experienced a

Table 8
Population of Province of Buenos Aires (Excluding the Capital),
Total Province (1797–1895) and 30 Counties
North of Salado River (1836–1895)

Year	Total population	Annual rate of growth (%)	Population, 30 counties north of the Salado	Annual rate of growth (%)
1797	32,168		—	
		2.39		—
1836	80,729		47,614	
		4.46		4.37
1854	177,060		102,855	
		3.96		3.08
1869	317,320		162,169	
		4.32		3.29
1881	526,581		239,259	
		4.20		1.62
1890	762,551		276,597	
		3.85		2.57
1895	921,168		314,090	

Sources:

For 1797 and 1836, Ernesto Maeder, *Evolución demográfica Argentina desde 1810 a 1869* (Buenos Aires, 1969), 32–34.

For 1854, *Registro estadístico del Estado de Buenos Aires*, 2° semestre 1854.

For 1869, *Primer censo de la República Argentina* (Buenos Aires, 1872).

For 1881, *Censo general de la Provincia de Buenos Aires, demográfico, agrícola, comercial, industrial* (Buenos Aires, 1883).

For 1890, *Censo general de la Provincia de Buenos Aires, 1890* (Buenos Aires, 1891).

For 1895, *Segundo censo de la República Argentina* (Buenos Aires, 1898).

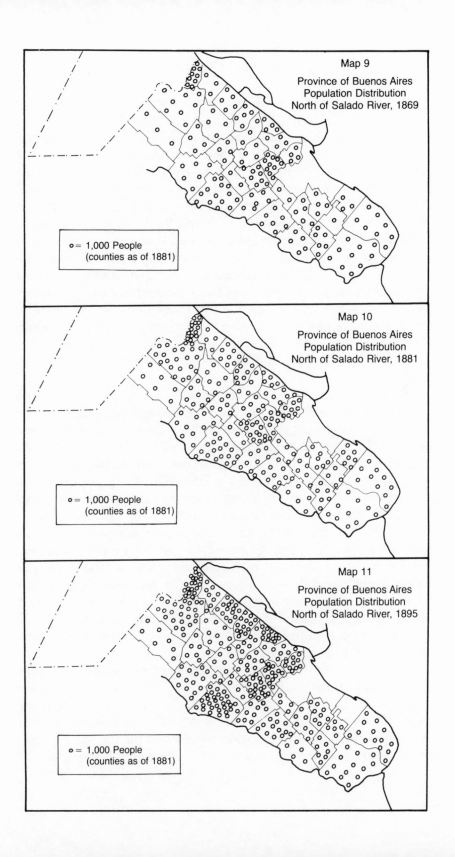

Map 9
Province of Buenos Aires
Population Distribution
North of Salado River, 1869

o = 1,000 People
(counties as of 1881)

Map 10
Province of Buenos Aires
Population Distribution
North of Salado River, 1881

o = 1,000 People
(counties as of 1881)

Map 11
Province of Buenos Aires
Population Distribution
North of Salado River, 1895

o = 1,000 People
(counties as of 1881)

more rapid growth than those of older settlement. Thus Pergamino and Rojas multiplied their population by more than five
times between 1854 and 1895, while Monte, Ranchos, and San
Antonio de Areco less than tripled their population in the same
period. Yet absolute figures may prove misleading, as the area
considered as belonging to a particular county varied in the different censuses, and therefore we have to pay attention to population densities, as shown in Maps 9–11. Counties of early
occupation and which started experimenting in colonization and
agriculture seem to have increased their population density at a
fast pace; Lujan, Chivilcoy, Mercedes, Pilar, and San Vicente being
amongst those with higher densities throughout the period, as
well as with the highest increase in those densities. In some of
these counties like Mercedes and Chivilcoy, as well as in others
like San Nicolás and Salto, the changes in density reflect both a
large increase in rural population and a very important growth
of urban centers.

Finally, we should point out that although there was a steady
increase in the population of the area during this period, its rate
of growth was lower than that for the province as a whole, and
it gradually represented a smaller proportion of the total. As new
territories to the south of the Salado were being opened, settlers
poured into that area, and if by 1854 58% of the population of
the province lived in the thirty counties selected north of that
river, by 1895 it had gone down to 34%.

(b) A very important factor in this population growth was immigration. By 1881, immigrants accounted for almost a fourth of
total population. Italians and Spaniards were the largest groups,
while French and British followed, with decreasing importance
as we come to the close of the century (see Table 9). As this subject
has been broadly studied by historians, I shall limit my analysis
to those aspects of immigration which are relevant to this work,
and those will be considered below.

(c) Throughout the period, the rural population was larger than
the urban population, but the latter increased in relative importance, reaching 36% of the total by 1881, north of the Salado River.
New towns emerged and old ones expanded as they became important centers in the provision of services for the surrounding
rural areas, and strategic points in the commercial, financial, and

Table 9
Foreign Population by Country of Origin, Province of Buenos Aires,
30 Counties North of Salado River, 1854, 1869, 1881, 1895

Countries of origin	1854[a]		1869		1881		1895	
	Number	%	Number	%	Number	%	Number	%
Spain	1,920	29	6,465	23	12,306	23	18,527	20
France	696	11	5,939	21	7,694	14	9,956	11
Britain	2,114	32	4,398	16	6,585	12	4,234	5
Italy	509	8	6,589	23	23,269	43	51,064	56
Germany, Switzerland, and Austria	193	3	1,143	4	1,655	3	2,742	3
North and South America	71[b]	1	2,076	7	2,855	5	3,513	4
Other countries	1,083	16	1,620	6	307	—	954	1
Total foreign	6,586	100	28,230	100	54,671	100	90,990	100

[a]Figures for 1854 are incomplete, and it is possible that the number of foreigners is underestimated.
[b]Includes only North Americans.
Source: Same as Table 8 for 1854, 1869, 1881, 1895.

transport networks which were being set up in connection with the new productive activities. Chivilcoy, Chacabuco, and Suipacha were among the *cabeceras de partido* to be created in this period (in 1854, 1865, and 1879 respectively), together with secondary towns which started to appear in the flat and previously almost uninhabited countryside. The role of these urban centers should be further explored.

(*d*) Figure 4 shows the composition of the population by sexes and age groups, north of the Salado. The effects of immigration are obvious: an expanding proportion of the people in the working age group of 16 to 50, and an increasing masculinity rate (see Table 10).

(*e*) The occupational structure of the population is considered only for the province as a whole, as there are no available data by counties. Table 11 shows that the potentially active population grows steadily during the period under consideration, but the recorded working population does not follow the same pattern, reaching its lowest points in 1854 and 1881, when only 47% and 58% of the potentially active population appears as employed. The high proportion of people in active ages with no recorded occupation probably reflects more the defective way in which information was collected than any significant fluctuation in the number of the employed. Moreover, these figures hide family labor. This generally occurs in rural areas, where the man of the house is registered as shepherd or agriculturalist, while his wife and children are considered as having no occupation, but in fact they perform part of the work the man is being paid for (see below).

Throughout the second half of the nineteenth century there was a clear predominance of rural activities, but after 1881 the number of those engaged in livestock raising diminishes in relation to total population, while that of agriculturalists increases steadily. The distribution for the rest of the occupations seems to remain more or less constant throughout the three censuses— excluding the very unreliable figures of 1854—with commercial activities showing an upward trend. The largest single group in the classification is that of the *jornaleros*, defined by the censuses as those having no established jobs, and which probably comprises mainly those who work in rural establishments as peones.

Figure 4

Province of Buenos Aires,
North of the Salado River, 1869, 1881 and 1895:
Population Structure by Sex and Age Groups

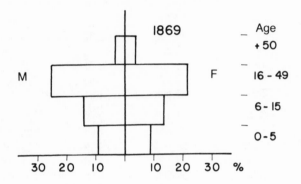

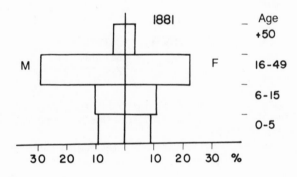

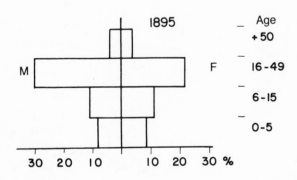

Source: Table 10

Table 10
Data for Figure 4

Year	Age groups	Males	% of total	Females	% of total
1869	0–5	15,186	9.37	14,682	9.06
	6–15	23,872	14.73	21,649	13.35
	16–50	42,167	26.01	34,950	21.56
	50+	5,673	3.50	3,930	3.42
	Total	86,898	53.61	75,211	46.39
1881	0–5	23,175	9.69	21,821	9.12
	6–14	27,844	11.64	25,705	10.74
	15–50	69,844	29.19	53,656	22.43
	50+	9,706	4.06	7,508	3.13
	Total	130,569	54.58	108,690	45.42
1895	0–5	26,739	8.54	25,886	8.27
	6–14	36,635	11.70	34,343	10.96
	15–50	93,003	30.65	68,402	21.84
	50+	14,697	4.69	10,476	3.35
	Total	171,074	55.58	139,107	44.42

Masculinity rate: 1869: 115.56%
1881: 120.17%
1895: 125.12%

Sources:
For 1869, *Primer Censo de la República Argentina* (Buenos Aires, 1872).
For 1881, *Censo general de la Provincia de Buenos Aires: Demográfico, agrícola, comercial, industrial* (Buenos Aires, 1883).
For 1895, *Segundo censo de la República Argentina* (Buenos Aires, 1898).

As we shall see, these jornaleros constitute an important part of the labor force employed by the estancias on temporary bases to perform occasional jobs.

Let us now analyze more closely the rural occupations as they developed between 1854 and 1895. A lack of homogeneity in the categories used by the different censuses leaves us with only three groups to consider: those engaged in agriculture on a permanent basis—whether they be owners of a farm, tenants, sharecroppers, or workers; those engaged permanently in livestock raising—whether they be estancieros, tenants, *aparceros*, puesteros, or shepherds; and the jornaleros, or laborers, who work on any type of rural establishment and may be hired on a permanent or temporary basis. Table 11 shows the evolution of these three groups throughout the four decades under consideration, and we can see how the proportion of jornaleros to the total working population remains quite constant, while *ganaderos* increased their relative numbers until 1881, later to lose ground to agriculturalists.

This is the general framework in which the particular population involved in the pastoral industry north of the Salado River develops. However, the statistical description of the structure of population in the province tells us very little about the process of formation of labor supply, a question to which I shall now turn.

The Supply of Labor

Population increased systematically, and it expanded into the rural areas, yet in the first decades of the pastoral era, the traditional shortage of hands seems to become only more acute. It was not simply a matter of population being scarce; perhaps the main problem was that potential workers found ways of making a living which did not require their permanent engagement in wage labor. A scattered population, large extensions of land, and herds of semiwild cattle with no legal or actual owners, and the presence of an open frontier gave the potential labor force a very anarchic character.

"The *vagos*, a plague in rich and fertile countries like ours . . . people the countryside . . . but [this evil] . . . will not disappear until civilization degrades the *chiripa* and the influx of immigrants stops the means of living without working."[10] For the rural settler

Table 11
Occupational Structure of Population in the Province of Buenos Aires, 1854, 1869, 1881, and 1895[a]

Occupations	1854 Number	1854 % of EAP	1854 % of total	1869 Number	1869 % of EAP	1869 % of total	1881 Number	1881 % of EAP	1881 % of total	1895 Number	1895 % of EAP	1895 % of total
I. Economically active population (EAP)[b]	51,775	100	41	134,389	100	71	181,580	100	57	346,800	100	61
1. *Primary sector*	15,620	30		32,935	24		53,380	29		82,677	24	
1.1 Cattle raising	9,586	19		20,357	15		31,637	17		35,525	10	
Estancieros/hacendados	—	—		(12,581)[e]	(9)[e]		(31,208)[e]	(17)[e]		(23,697)[e]	(7)[e]	
Qualified workers	—	—		(7,776)[e]	(6)[e]		(429)[e]	(—)[e]		(11,828)[e]	(3)[e]	
1.2 Agriculture	5,764	11		10,919	8		15,828	9		45,694	13	
1.3 Others in primary sector	—	—		1,659	1		5,915	3		1,458	1	
2. *Secondary sector: crafts/manufacture*	2,257	5		17,997	14		20,063	11		44,522	12	
3. *Tertiary sector*	2,891	5		31,839	24		37,637	21		84,734	24	
3.1 Commerce/finance	2,130	4		7,233	5		12,504	7		22,762	7	
3.2 Government[c]	761	1		2,367	2		3,483	2		7,069	2	
3.3 Transport/commun.	—	—		4,163	3		3,914	2		11,880	3	
3.4 Arts/professions	—	—		1,420	1		2,506	2		5,767	2	
3.5 Domestic service	—	—		16,656	13		15,230	8		37,256	10	
4. *Sector unknown*	31,007	60		51,618	38		70,500	39		134,867	39	
4.1 Clerks	1,923	4		2,710	2		2,985	2		10,116	3	
4.2 Jornaleros/peons	20,313	39		48,825	36		67,429	37		123,467	35	
4.3 Others	8,771	17		83	—		85			1,284	1	
II. Nonactive population[b]	—			1,969		1	5,208		2	10,316		2
III. Occupation unknown[d]	74,307		59	52,507		28	131,160		41	212,621		37
Total of potentially active population	126,082		100	188,865		100	317,948		100	569,737		100

[a]The occupational tables of the censuses of 1881 and 1895 include only population over a certain age (15 for 1881, 14 for 1895). The census of 1869 does not mention limit, but retrospective information included in the census of 1895 refers to the 1869 census as including only population over 14 years old. The provincial census of 1854 does not mention age limit.

[b]Economically active population (EAP) is taken to mean the population that appears with occupation in the censuses, except for occupations that do not imply participation in economic activities (for example, rentiers, students, nuns, etc.). Technically, the latter are added up in the category of nonactive population.

[c]Includes the military.

[d]The censuses of 1881 and 1895 include a category of *sin profesión* (without occupation) or *profesión desconocida* (unknown occupation), but the census of 1869 brings no such information, and the census of 1854 has greatly underestimated the number of those with no recorded occupation. Thus, in the case of 1881 and 1895, I have included in the table the figures of the respective censuses, although the total population over 15 years of age results in 317,948, while in the tables by age groups, this population only adds up to 312,575. For 1854 and 1869 I have calculated the number of those with unknown occupation by subtracting the total of active and nonactive from the potentially active (all those over 14).

[e]Figures in parentheses represent portions of figures for category 1.1.

Source: Same as Table 8 for 1854, 1869, 1881, 1895.

it was not necessary to *work* for a living and therefore only occasionally did he sell his labor in the market. "He is a man of little
needs," says Marmier, "and he is contented with the scarce resources he owns, with the few pesos he earns as a *peon*."[11] However, these men were not peasants, integrated to a natural or
subsistence economy. "As a butcher of stolen cattle, as a thief
related to the Indian or to other *hacendados*, as a hunter of *bichos*
(from otters to ostriches and foxes) who does not mind hunting
cows and stealing horses when opportunity offers,"[12] it was the
possibility of establishing a direct connection with the existing
commercial networks and of selling the produce of his own "harvest" that gave the rural settler access to the money he needed
for his living.

Under these circumstances, uncertainty reigned in the market,
as the labor supply was almost unpredictable and could not be
expected to respond to variations in demand. These variations
were quite acute, as the estancias were organized to depend as
little as possible on permanent labor and thus relied heavily on
seasonal and casual workers. Supply seldom met demand, and
thus followed the traditional scarcity of labor in Buenos Aires
province, a scarcity made more acute by the unprecedented expansion inaugurated in mid-century.[13]

At this point, the creation of a stable and predictable supply of
labor to meet the demands posed by the expanding economy
became the main object of this phase in the process of organization
of the labor market. There were two principal sources for the
creation of this supply in Buenos Aires. In the first place, those
native workers who, though occasionally participating in the market, had enjoyed alternative ways of subsistence other than wage
labor, now were driven into the market as a regular and reliable
work force. Second, immigrants, who had arrived in large numbers since mid-century, soon became the main source of labor
supply.[14]

Native Workers. Since the first decades of the nineteenth century,
the state authorities and the propertied classes of Buenos Aires
had been concerned by the lack of labor discipline in the rural
population and with the difficulties in exercising any sort of social
control over it.[15] The estancieros were faced with several prob-

lems. In the first place, as we have seen, there was the shortage of labor, partly a consequence of the possibilities men had to secure life by methods other than wage labor. It was not only a question of hiring laborers, but, above all, of being able to retain them, preventing their leaving their place whenever they saw fit, quite often before concluding the job they had been hired for. The estancieros complained because peons deserted "for one, two or more days the job that has been entrusted to them"[16] and because "they frequently leave."[17] In order to attract workers, employers advanced cash to them, and when the government tried to stop this practice, the claims of the estancieros were most eloquent: "It is useless to restrain or forbid those advances the employers want to make to their workers: the scarcity of hands compels them to do that even at the chance of losing everything."[18] In fact, advancing their wages to them was an incentive offered to the workers in order to attract them to the job rather than a means to tie them through indebtedness.[19]

Second, the usual practice of occupying land which had no apparent owner and stealing cattle for personal consumption or for sale became a threat to private property. Moreover, as produce increased in value, the loss of a horse, a few hides, or the wool of a flock became a significant loss in terms of income, which the estancieros wished to avoid.

In view of these problems, the opinion of the estancieros was unanimous: They clamored for measures to control and discipline the population of the countryside. On account of the drafting of the Rural Code, on two occasions—1856 and 1863—the estancieros were asked their opinion on different matters which were to be considered in the new legislation basically intended to impose law and order in the province. In their answers, compiled in one volume, two kinds of suggestions prevail: those directed toward preventing rural settlers from having alternative means of subsistence other than wage labor and those referring to specific measures of control and repression of the population.

Among the first type, estancieros insisted on the need to stop the practice of allowing those "with no means of subsistence [that is, no cattle and no real estate that might produce rent] to settle in the borders of an estate . . . in a small hut where they stay with more or less family . . . working maybe no more than three months a year . . . staying for years and years, though the opinion of the neighbors is that they live regularly on what they steal."[20] Relocation to frontier areas or local towns is recommended in these cases.

There is a shared feeling against traveling salesmen who "generally buy stolen goods" and against *pulperías* (local stores), "the greatest plague of the country."[21] The opinion of most estancieros is that such practices as hunting ostriches, gathering straw, grass, or firewood, and even picking up bones should be forbidden because these resources "belong . . . to the owner of the estate" who may "exploit them as he wishes."[22] To stop horse stealing—a common practice of the day—several measures of different scope are proposed, from controlling tradesmen to forbidding the use of the *bota de potro* (boots made by the rural settlers with the skin of a horse leg).

Regarding actual measures of control of the rural population, estancieros show their preference toward legal papers, such as the job certificate (*papeleta de conchabo*) and the passport to move about the province, whose main purpose was to discourage mobility (see below). They also show deep concern regarding the enforcement of such measures, and include proposals to increase the efficiency of police forces in the rural areas.

There is one aspect, however, in which the opinion of the estancieros is divided, and that is the levy system. At this point, the demands of the estancieros overlap the needs of the state. For the state, though committed to the expansion of the export economy, it was a question of not only meeting the demands of the productive sector but also of exerting social control over the population of the province. Law and order were key words for the local administration, and the successive governments in this period improved the legal instruments Rosas had already counted upon—among other more direct means—in his successful campaigns to gain control of the countryside. Furthermore, for those who were or expected to be in power, this legislation, and particularly the means devised to enforce it, became a decisive instrument to achieve political control.

These measures became even more important at the time of organizing the provincial army. Since the days of the Revolution, the Indian peril, interprovincial conflicts, and international wars had compelled the local authorities to keep a standing army, whose members were recruited mainly from among the rural population. Thousands of adult males—potential workers—were sent to the front every year, greatly affecting the supply of labor.

In this context of competing demand for the same men, legislation on vagos and malentretenidos has been seen as a tentative solution for both problems, i.e., controlling the labor force *and* providing men for the army.[23] This was done by defining any man with no legal property and no established job, with no passport or *papeleta*, as a vago, liable of being arrested by the local authorities and sent to the army for a few years to serve as soldier. By intimidating the worker and punishing the transgressor, the authorities hoped both to favor the creation of a reliable work force and to ensure the supply of men for the army.

The main stages of this legislation are well known.[24] The *Bando Oliden* in 1815 had defined any propertyless person as belonging to the class of servants, with the obligation to find employment and to carry with him a certificate (papeleta) signed by his employer and by the *Juez de Paz* and renewed every three months. If found without such a certificate, the man would be classified as vago and sent to the army for a period of five years, and if unfit for that service he could be forced to serve under any employer.

More legislation to the same effect was passed in the 1820s and 1830s, and the Rosas regime was characterized by an effective application of these measures. The choice was wage labor or the front, and frequently the decision was not even in the hands of the individual potential worker, but in those of estancieros and justices of the peace, who at any time could accuse him of *vagancia* and send him to serve in the army. These arbitrary measures were aggravated when the government ordered the recruitment of a definite number of soldiers for the front, for actually the local commander had power to choose whom to send, what estancieros to favor by sparing their men, and which to "tax," as MacCann put it in 1848.[25]

During these decades, moreover, passports were required to travel from one county to another, a requirement which was abolished after the fall of Rosas in 1852, only to be re-established the following year by Urquiza. In 1853, 1855, and 1858 new measures were passed for recruiting vagos and malentretenidos. Finally the Rural Code was approved in 1865, legally regulating the rights of property, the relationship between landowners, the organization of rural police, and the relationship between employer and worker, and thus defining the rights and obligations of master and peon, as we shall see below.[26]

For the Rural Code *vago* was "anyone with no established residence or known means of subsistence, breaking moral laws as a consequence of bad conduct or usual vices."[27] He was to be judged by a local jury of neighbors and if found guilty, he could be sent to serve in the army for three years or condemned to forced labor for one year. In the same year of 1865, a regulation was passed exempting managers and supervisors (mayordomos and capataces) from the levies, and giving the estanciero the possibility of replacing any of his men by someone else, paid to act as a substitute.

Although this legislation was meant to meet the requirements both of the productive sectors and of the state, it was soon to prove conflicting. Already in the late 1850s there were different opinions as to the best way to attract potential workers to the estancias, and some estancieros and other observers found the levies only made matters worse in terms of labor supply by prompting men to leave the province to avoid recruitment. Later on, voices of protest rose louder, condemning the arbitrary levies, suggesting fairer methods of recruitment, denouncing the miseries of life at the front and the injustices suffered by gauchos and rural workers.[28] Protests also arose from among those who employed rural laborers because levies frequently left them short of the necessary hands in their establishments, and increasingly the men were recruited not for guarding the frontier, but for fighting wars in distant territories.

Nevertheless, in the 1860s not a few of the estancieros were still in favor of the levy system. So much so, in fact, that on the occasion of the War with Paraguay, we find *The Brazil and River Plate Mail* praising the effects of the conflict on the pastoral industry, saying: "The present war is more advantageous for them [the sheep farmers] than otherwise, inasmuch as it rids the camp of a number of vagrant gauchos, whose only visible means of existence was stealing horses."[29]

But soon this system that had been relatively efficient by mid-century started to show too many flaws. As the social and economic structure of the rural areas changed, new problems and contradictions arose. Thus, for example, if on one side it was important to stop peons from leaving their jobs before their contract was over (and in this sense measures like the papeleta and the passport were efficient), on the other side, the pastoral economy required increasing numbers of seasonal and casual laborers and therefore it benefited from free mobility of the labor force, and in this sense the passport became an obstacle. But not only the text of the laws was problematic. The instruments devised for enforcement brought about conflicts of jurisdiction and power at the local level, and the relationship between *comisarios, jueces de paz, comandantes* and estancieros was seldom smooth.

Gradually, many of these measures were abolished. In 1870 the Rural Code was reformed to suppress the articles on vagos and in 1873 the passport requirement was abolished. Yet it would be only in the late 1880s that a system of lotteries was established for the recruitment of those who were to serve in the army.

In this way, therefore, throughout the period under consideration the propertyless native population of the countryside was compelled to work in estancias and farms or to go to the front. Of those who chose or were forced to choose the first alternative, many found their way into the pastoral industry, where they worked together with laborers who had a very different origin; namely, the immigrants.

Immigrant Workers. In the second half of the nineteenth century, immigrants became the main source of wage labor in the province of Buenos Aires, contributing greatly to the creation of an expanding labor supply. They were to be found in most branches of the local economy, not least in the pastoral industry.

Starting in the late 1840s, Buenos Aires had witnessed the arrival of increasing numbers of Irish, Scottish, and Basque settlers, who somehow had found their way to the shores of the River Plate, attracted by uncertain but nevertheless promising perspectives. They happened to arrive at the right place at the moment when the expanding pastoral sector found in them the skilled and reliable labor force it was so desperately in need of.[30]

It was nonnatives with names like Sheridan, Harrat, Hannah, and Stegmann who had developed an interest in sheep breeding in the first place. After starting the business, they were followed at first by some pioneer Argentine cattle estancieros, and later by the bulk of the ganaderos of the province, as well as by new investors in the rural field. These foreign estancieros from the very start employed immigrants as managers and supervisors, and definitely showed a preference toward foreign shepherds. These newly arrived Irishmen, Scots, and Basques proved the ideal type for sheep raising. Mostly they were skilled, and when not, they rapidly acquired the abilities necessary to carry on any type of work. They were immigrants in search of social and economic improvement, and were ready to perform whatever job was required of them. Most of them were used to peasant family labor, which—as we shall see—would prove very efficient in the context of a sheep farm or an estancia. When workers belonged to the same nationality as the owner of the establishment, as in the case of many Irishmen and Scots, bonds of patronage would arise between worker and employer, not infrequently reinforced by religious and cultural ties. Furthermore, being foreigners, these workers were not subject to the military levies which were so frequent in the period and area under study.[31]

As early as 1848, MacCann observed: "The banks of the river in the neighborhood of Chascomus are very densely peopled by British subjects, chiefly Irish, all employed on sheep-farms."[32] A decade later Chaubet would say, "The Irish form 3/4 of the English immigration . . . most of them become shepherds,"[33] and Hournon: "It is mostly Basques and Bearnais who have gone to represent France in the Río de la Plata. They are wanted, looked for, even preferred to the gauchos by the estancieros."[34]

"The British population in the Argentine Republic is calculated at 32,000, of which number 28,000 are Irish, forming about 5,000 families, and residing for the most part in the country, where they occupy themselves in rural pursuits and the tending of sheep," stated a report published by the Parliamentary Papers in 1866.[35] And President Mitre recalled in 1870, "Who has fostered sheep raising, that has brought welfare to our countryside and multiplied our commercial transactions? Sheep can be said to be the core around which Irish immigration has developed."[36]

Even in the 1880s and 1890s Irish and Basques would still be mentioned as the most successful in sheep farming. "The Basques, from Southern France and Spain are specially welcome here. . . . They frequently arrive with their families and as a rule remain, working chiefly at agricultural pursuits and in sheep and cattle farms. . . . Amongst the most successful—if not the most successful in comparison to their numbers—of all immigrants in this country have been the Irish. . . . They generally arrived with little or no capital and set to work and save as shepherds."[37]

Finally, in 1892, a report mentioned: "Native and foreign labor take different directions. Generally speaking, it may be said that, amongst the lower classes, stock-raising is in the hands of the natives. . . . Cattle-farming (as opposed to sheep-farming) is almost exclusively in native hands. Sheep-farming, however, is often undertaken by foreigners, the Irish and Scotch being prominent in this respect."[38]

For the counties north of the Salado River, we can see the number of foreign settlers in Table 9. But unfortunately there is no way of knowing their occupational distribution, or whether they were established in the rural or urban areas. It is also hard to assess their real weight in the population of each county, as sons and daughters of immigrants born in Argentina are classified as natives, although they shared many of the social characteristics of their parents.

Italian and Spanish immigrants predominated in the area under study, but contemporary observers point out that these groups were massively dedicated to agricultural or commercial activities. Frenchmen in general had varied trades, but concentrated in arts and crafts; while Basques worked in saladeros, brick factories, dairy farms, and sheep raising. Germans and Swiss were concentrated in colonies and small rural towns.[39] It is particularly the Irish and Scots who were, as we have seen, dedicated to sheep raising.[40] In fact, the counties north of the Salado show a higher percentage of British-born than the province as a whole throughout the period under study, while the opposite is true for the other nationals.

How were immigrants attracted to the pastoral industry? Since a very early date after Independence, immigration and colonization proposals had been in the agenda of successive provincial governments. Argentine ministers and envoys to Europe kept advertising the extraordinary conditions the country offered to foreign settlers, but few measures were actually taken to promote organized immigration to the province of Buenos Aires under official protection, and most of them ended in failure.[41]

Thus, most of the immigration in the early decades of sheep expansion was spontaneous, resulting generally from more or less personal connections of the settlers-to-be with residents in the River Plate. In the case of Irish and Scottish immigrants, the early development of an organized community in Argentina seems to have acted as an important nexus between the newcomers and the country they chose to settle in. Influential personalities, like Father Fahy in the Irish case, were often decisive in attracting settlers from the home country.

Reasons such as good climate and excellent predisposition toward foreigners, healthy atmosphere and good-natured people, fertility of the soil, and civil and religious liberty were advanced in favor of Buenos Aires. There were of course counter arguments, and many an official report in Great Britain, France, and the United States tended to discourage their people from migrating to the River Plate.[42] But those who found their way to Buenos Aires, and who decided to try their luck in the expanding pastoral industry, had one good argument: the high wages paid to foreign workers and the favorable terms of engagement to be found in that field, particularly during the first two or even three decades of expansion.

Foreign workers were not the only labor to be employed in sheep raising, however, and though Irish, Scots, and Basques were preferred by the estancieros to shepherd their flocks and manage their establishments, most of the peons and a considerable number of puesteros were natives. Thus, both natives and immigrants contributed to the supply of labor in the pastoral sector where they were employed under the conditions we shall now turn to analyze.

Forms of Engagement

Wage labor was predominant throughout the period under consideration in the rural areas of the province, north of the Salado

River. Yet, it appeared under different forms and coexisted with other types of engagement, reflecting the various ways in which capital related to labor. Relationships between employer and employee can be understood only within the context of the social system of which they are a part, and therefore my attempt to portray at this stage the conditions of the labor force will necessarily be partial and incomplete. Future chapters on the process of production, circulation, and the characteristics of capital accumulation will enrich the limited picture that may result from the present description, although a complete study of social classes is out of the scope of this work.

From the point of view of the way in which surplus value was extracted from labor, three general forms of engagement were to be found in those forty years during which sheep raising and wool production were predominant in the area under study: These were wage labor strictly speaking, *aparcería*, and family labor.

Wage Labor. The more widespread form of engagement was that of wage labor strictly speaking, in which the worker was hired to perform a job—either temporary or permanent—and was paid a wage generally part in cash and part in kind—mainly food and lodgings. Different types of jobs were performed under the wage system, from the more simple tasks of the peon to the highest supervising jobs of capataces and mayordomos.

The peon was the field laborer hired by estancieros or farmers for the most varied kinds of jobs. The Rural Code of 1865 defined him and his employer as follows: "Rural master is he who hires the services of any person in the benefit of his rural property, and rural peon is he who renders them in exchange for a certain price or wage."[43] According to the Code all workers, except day laborers, had to be hired by written contract, establishing the conditions of the engagement. The peon was to live in his master's quarters, and to work the days and hours fixed by the contract, except Sundays which he was to have free, though not during the shearing season. If disobedient, lazy, or "vicious" he could be fired by the employer, but the contract could not otherwise be broken without proper "justification."[44] These were the legal conditions of engagement for the peones, although their observance was far from being general.

Peones were hired on a permanent basis to help in the general work of a sheep farm or an estancia, but more often they were called for temporary jobs. Sheep raising was a seasonal activity which required extra hands at specific times during the year. Thus, for operations such as lambing, the cutting of the males, and especially shearing, estancias and farms had to hire temporary peons to help the regular staff with the additional tasks. If it is quite surprising to see the number of people registered as jornaleros having no established jobs in the censuses of 1854, 1869, 1881, and 1895 (see Table 11), it is more so to find out that in 1888 the personnel employed on a temporary basis by rural establishments in the province was considerably more numerous than that which was hired permanently.[45] Although these temporary peones most probably worked for different employers throughout the year, and therefore the figure of 219,500, shown by the Census of 1888, does not represent the number of peones but the number of jobs performed by them, it nevertheless suggests a large number of workers changing employers and jobs several times during the year. Although these figures refer to agriculture as well as to stock raising, they are indicative of a situation that was common to both fields of production.

Table 7 includes an estimate of the total number of permanent and seasonal workers *required* by the pastoral sector throughout the period under study. The figures were calculated on the basis of the information available on total wool production for the province for the different years and on average productivity of labor for the whole period, and therefore they do not refer to the workers actually employed in the pastoral sector, but only to those required by it. It is obvious that the resulting figures should be taken only as very rough estimates which may shed some light on the relationship between permanent and seasonal labor requirements. In this sense, and although the figures refer to all hands needed by the pastoral sector—whether they be wage laborers, sharecroppers, or even farmers in charge of their own flocks makes no difference in this estimate—it is quite clear that the seasonal rise in demand was so sharp that it surely required the employment of extra hands—wage laborers engaged for the shearing season.

This was the high season in the pastoral industry. All human resources available in the area were used during that period, and there is evidence of seasonal migrants from neighboring provinces coming to Buenos Aires province for the occasion.[46] The shearing took place within the head station of each estancia, and all the labor force of the establishment was engaged, besides the additional hired labor. On sheep farms, the whole family and frequently neighboring farmers participated in the shearing, yet extra hands were always hired during that time of the year.

Those in charge of the actual shearing were paid by the number of sheep shorn, and workers who caught the sheep from the flock (*agarradores*) and those who collected the fleeces and made bundles out of them (*atadores*) could be paid by day or by piece. Food and lodging were provided for them while there was work to be done. It is interesting to note that frequently women came with their men to the establishments, and they as well were employed as shearers. Contemporary observers have left colorful descriptions of the *esquila*, the most complete description being that of Gibson, and although they are somewhat lengthy, his paragraphs are well worth quoting:

This [the shearing], the most important occurrence in the annual history of the sheep-farm, generally commences in the first fortnight of October, and should be terminated before the first fortnight of December, before the grass seeds and burrs have begun to come away and get into the fleece. It is not customary to wash the sheep before shearing. . . .

[Sheep are shorn in] a yard. . . . The sheep are driven into the yard and caught by men whose whole duty is to attend to this department. They tie three legs of the sheep together, and place the animals conveniently near the shearers, the legs being tied with a thong made of teased rags or sheepskin. These men are paid from 12*d* to 15*d* per hundred, and one catcher is supposed to be sufficient to supply ten shearers. The shearers are paid from 7*s*. to 10*s*. per hundred. . . .

Each shearer should have a small pen, capable of holding from 15 to 20 shorn sheep, and as he finishes the shearing of an animal he turns it into this small enclosure. When the pen is full, he calls the overseer, who revises [sic] the shorn animals, and if they are carefully clipped, free of cuts, and all the leg locks neatly removed, he counts them out, giving the shearer tokens for the number. If he finds any badly shorn he reprimands the man; if the offence is repeated he discounts the badly shorn animals; and if carelessness continues he dismisses the shearer. I may here remark that the shearers and other men employed at the work—the former being chiefly natives, and the latter Spaniards, Basques and Italians—are all obedient and attentive to their work, and there has been no experience of the strikes and troubles created by the same class of men in Australia.

The fleeces are collected by boys, who earn a monthly wage of from £1:10s. to £2 and carried by them to the tying tables. . . . The wool-tyers, who also bag up the wool, are generally contracted for by the piece. . . . If the fleeces are to be classified for the home market, they are thrown by the wool-tyers to the centre of the classifying floor, where the sorter separates them into their corresponding divisions. If the wool is for sale in the Buenos Aires market or for delivery to a local buyer, the fleeces are bagged up immediately and stored away. . . .

It is usual to employ as many of the permanent labourers of the estate as possible, but of course these are not sufficient, and others are hired from elsewhere for the shearing season. All alike get three meals per diem, viz one at 8 a.m., one at mid-day, and one at sunset. The shearers are usually supplied with a ration of 1 lb. of *yerba* . . . per week and the cooking is done for them by a man hired by the sheep-breeder. Payment can either be made by cheques upon some neighbouring store, or in cash. It is a rule only to pay on Saturdays, and never to allow a labourer to withdraw more than 50% of his earnings until the whole of the shearing is concluded and the hands paid off.

Shearing is commenced as early in the day as possible. . . . Half an hour is allowed for breakfast at 8 a.m., and an hour at mid-day, but with these two exceptions the work continues steadily from sun-up to sun-down.[47]

With little differences, authors like Vicuña Mackenna for the 1850s, Latham for the 1860s, McLeod for the 1870s, Emile Daireaux and José Hernández for the 1880s, and Godofredo Daireaux for the last years of the century have portrayed the events of the shearing season in the estancias of the province of Buenos Aires.[48]

What happened to the temporary laborers once the season was over? There are no references to such a problem in any of the consulted sources, but there are grounds to infer that there was enough work to do in estancias and farms to employ large numbers of temporary workers during the summer. It is Gibson again who says, "In special seasons, such as shearing, dipping for scab, etc., the day laborer is paid from 3s. to 4s. per diem, including mid-day meal, and sometimes both mid-day and evening meal and lodging. Specially hard work, such as digging of wells is paid at a higher rate, say from 5s. to 6s. per diem. When such work can be paid for by the piece, it is generally preferable to do so."[49]

The winter, however, was another matter, and if large establishments might nevertheless require some extra hands to dig a new well or repair an old fence, most smaller estancias and farms probably dispensed with temporaries during those months of relative inactivity. Those workers who had come from provinces like Córdoba, Santa Fe, and Santiago del Estero probably went back to their homes at this time of the year, and other workers would just ride from estancia to estancia trying to find a job.[50] Sometimes, they would be hired for a few days, at other times they would be caught by the local authorities and sent to the front as vagos. There is very little evidence of the fate of these peons, and further research should be done on this subject.

The life of permanent peones seems to have been less hazardous, though hardly any better. According to Gibson: "A general laborer whose duty is to dig ditches, work in the wool-shed and generally assist as a foot-workman at the head station, is paid from £2 to £3 a month. In addition to this he is found in both house room and board. His food is cooked for him, and he usually gets one meal at mid-day and another at sun-down, as well as a cup of tea, cold meat, or something similar before sun-up. Sometimes, he gets bread and tea rations instead of being supplied with them in the common kitchen. These consist of 3 lbs. of 'camp' biscuit, 1 lb. of sugar and 2½ oz. of tea, or if he prefers, 1 lb. of Paraguayan tea per week."[51]

Among the permanent workers, the puestero was in a slightly higher category than the peon. We have briefly seen what the main tasks of these shepherds were, and many contemporary testimonies portray certain details regarding their way of life, their duties, miseries, and compensations.

Though it is not too difficult to portray the way of life of shepherds, it is almost impossible to find out exactly how many there were or where had they come from. The censuses of 1869 and 1895 register 7,529 and 10,444 puesteros and *pastores* respectively for the whole province, but there is no way of assessing the reliability of the figures, as the categories are not clearly defined and could include herdsmen and aparceros, as well as shepherds who are hired as puesteros. Argentines, Irishmen, Scots, and Basques seem to have been predominant among shepherds on wages in the province of Buenos Aires, north of the Salado River, but again no precise figures are available. Thus, for example, here is Latham's description:

> On different parts of an estancia there are erected the huts of the shepherds, with their "corrales," called "puestos" or stations, which, with a certain extent of ground, are allotted to the different shepherds or "puesteros" for the run of the flock under their care. For the maintenance of the "puesteros" the meat of a wether (capon) or aged ewe is allowed between three or four, the skin, tallow, and grease being set apart for the proprietor and collected periodically. The "puestero" provides his own "yerba" . . . and sugar, and, if he indulges in the luxury, his own biscuit and salt. Fuel he provides for himself from the stems of the hard "thistle," "bisnaga," or the excrement of the sheep. . . . He makes his fire either in the open air in the center of his hut or under a shed outside; he runs a long spit, "asador," through the meat and sticks it in the ground with an inclination over the fire. There are many, however, chiefly foreigners, who show more refinement, and possess a table and chairs, frying pan and saucepans, plates, knives, and forks, and will set a benighted traveller . . . down to a good stew, with rice and eggs, in addition to the customary sweet roast or "asado," laid out on a clean table cloth, a cup of genuine Congou or Souchong, with ewe's milk, and the never failing "drop"of spirit . . . and last, and best of all, a hearty English, Irish or Scottish welcome.[52]

Emile Daireaux and Estanislao Zeballos's descriptions for the 1880s and Gibson's for the early 1890s differ little from Latham's account.[53]

Within the estancia order, the capataces (overseers) were in charge of directing the work of permanent and temporary peons, as well as of controlling the shepherds and acting as overseers of the work done in the establishment. Their task will be more clearly understood when describing the organization of the units of production. Nevertheless, we could point out here that they were generally paid in wages, and that their living conditions were only slightly—if any at all—better than those of the rest of the wage laborers employed in the estancias.

The situation was quite different for the mayordomo (manager), who was the personal representative of the estanciero and had control over everything that went on in the establishment. It is hard to generalize about their job or living conditions, as those depended entirely on the relationship established between manager and owner. It was a highly specialized job, generally very well rewarded, and not a few of those who worked for a number of years managing an estancia in due time acquired their own land and became themselves prosperous sheep farmers.[54] Immigrants were preferred for this post, and most of the pioneer sheep breeders were very careful to choose Irish, Scottish, or German settlers to supervise their establishments. These were generally paid in wages, which were frequently complemented by a small share of the profits.[55]

Mayordomos and capataces represented only a small proportion, however, of the labor force employed in rural Buenos Aires. The bulk of the wage laborers were temporary workers, permanent peons, and puesteros, whose wages seem to have followed a parallel path during the second half of the nineteenth century. Table 12 includes data collected from different sources regarding wages in the period and area under study, considered in peso moneda corriente (m/c) and *oro*. Although these data are not precise, homogeneous, or regular enough to allow for the construction of a series, where we could analyze the repercussions of the different cycles of the pastoral industry in wages, they do provide an indication of the trend followed by them during the period under study.

I have chosen to express figures in golden pesos to make the table comparable to others included in this work, although it is clear that figures in golden pesos do not necessarily reflect the purchasing power of those wages. I have also included figures in paper currency, as they were received by the laborers. At the same time, it is necessary to take into account that these figures represent only part of the wage—that which was paid in cash—and that workers were also paid in kind (food and board). The actual

Table 12
Rural Wages in the Province of Buenos Aires, 1842 to 1895
(in Golden Pesos and Current Paper Pesos)

	Peones and puesteros (per month with board)		Peones (per day)		Overseers (per month with board)	
Year	$ m/c	$ oro	$ m/c	$ oro	$ m/c	$ oro
1842	55–100	3.3–5.9				
1848	100–150	4.7–7.0	20–25	0.9–1.2		
1850	50–100	3.3–6.7	5–10	0.3–0.7		
1854		11.6–15.6		0.5–1.0		
1855	150–250	7.0–12.0	8–12	0.4–0.6		
1858	150	6.8	20–30	0.9–1.4	200	9.0
1860		10.4		0.8		
1862		10.0				
1864		12.5–15.0				
1865	250–300	9.5–11.4	21–50	0.8–1.9		
1866		12.1–16.8		1.0–1.3		
1867	230–350	10.0–16.8		1.0–1.3		
1868	200–350	8.2–14.4	20–30	0.8–1.3	300–350	12.3–14.4
1869	250–350	10.3–14.4	20–30	0.8–1.3	300–350	12.3–14.4
1870	300–500	12.4–20.6	10–20	0.4–0.8	400–800	16.5–32.9
1871		10–12		1.2		
1872		12.6–21.0				
1875		12.5–20.8		0.8		
1876	416	14.2			600	20.6
1879	400–416	12.9–13.4			600	19.3
1880		13.3		0.8		
1884		14–24		1.0–1.8		
1885		12.5–14.6		1.0		
1886		14.2				20
1888	16.0–18.2	10.8–12.3				
1889		12–16				20
1890	20–35	7.8–13.6				
1891	20–25	5.3–6.7				
1892/3	15–45	5–15		0.7–1.0		20
1895	20	5.8	2.0	0.6		

Sources
1842 *Archivo Senillosa*, AGN, Sala 7, 2-6-13, "Estado de gastos que se an echo en el Establecimiento del Arroyo Chico del Sr. Don Felipe Sanillosa, 1° marzo y 1° junio 1842."
1848 W. MacCann, *Two Thousand Miles' Ride Through the Argentine Provinces* (2 vols., London, 1853), vol. 1, 25.
1850 *Sucesión Sheridan*, AGN, Sucesiones, No. 8184.
1854 A. Brougnes, *Cuestiones financiera y económicas de la República Argentina* (Buenos

Table 12 (continued)

Mayordomos (per month with board)		Servants and cooks (per month with board)		Shearing squads (per 100 sheep)	
$ m/c	$ oro	$ m/c	$ oro	$ m/c	$ oro
			11.6–15.6	57	2.9
					1.0
					1.0
			4.2		1.3
					3.1
		130–300	5.0–11.4		1.4
			10.0–20.6		
			11.6–20.6	40	1.7
800–1000	32.9–41.2	150–300	6.1–12.2	50	2.1
800–1000	32.9–41.2	150–300	6.2–12.2	50	2.1
		120–150	5.0–6.2		1.5
			12.6–25.0		1.7
750	25.7	340	11.7		
750	24.1				
			6.0		1.7
			12–30		
			6.2		1.9
	27.0				
			10.0		
		35–60	13.6		
					1.8
				3	0.9

Aires, 1863), 129; *Archivo Senillosa,* AGN, Sala 7, 2-6-13, Letter from Enrique Gubba to Don Felipe Senillosa, 9 Dic., 1854.

1855 B. Vicuña Mackenna, *La Argentina en 1855* (Buenos Aires, 1936), 136; *Registro estadístico del Estado de Buenos Aires,* 1st semester of 1855, 60.

1858 *Sucesión Sueldo y Gomez,* AGN, Sucesiones, No. 8184.

1860 Francisco Seguí, "Investigación parlamentaria sobre agricultura, ganadería, industrias derivadas y colonización," *Informe del Comisario Sr. Ingeniero D . . . Anexo B, Pcia. de Buenos Aires* (Buenos Aires, 1898), 67.

(continued next page)

Sources for Table 12 (continued)

1862 St. James' Magazine, 21 (1867–68), 299.

1864 The Brazil and River Plate Mail (BRPM), 21 June 1864, 321; Thomas Hutchinson, Buenos Ayres and Argentine Gleanings (London, 1865), 307.

1865 Seguí, op. cit., 67; Seymour, op. cit., 38–39; BRPM, 22 Feb. 1865; Archives des Affaires Etrangères (France), Correspondance commerciale, Buenos Aires (consular), vol. 7, 1865–67, 16–17, 12 March 1865; Sucesión Samuel Wheeler, AGN, Sucesiones No. 8760.

1866 Parliamentary Papers, Commercial Reports (b) Embassy and Legation, vol. LXIX, 1867, 317–318.

1867 Wilfrid Latham, The States of the River Plate (London, 1868), 32; BRPM, 6 April 1867; L. A. Dillon, Twelve Months Tour in Brazil and the River Plate with Notes on Sheep Farming (Manchester, 1867), 27; Anales de la Sociedad Rural Argentina, vol. 1, 1867, 75–76; Archives des Affaires Etrangères (France), Correspondance Commerciale, Buenos Aires (consular), vol. 7, 1865–67, 333–334, 12 Feb. 1867; Mathew to Lord Stanley, M.P. No. 5, Buenos Aires, 19 Jun. 1867, F.O. 6/267.

1868 W. Hadfield, Brazil and the River Plate in 1868 (London, 1869), 267–268; Sucesión Ricardo Newton, AGN, Sucesiones No. 7217.

1869 M. G. and E. T. Mulhall, Handbook of the River Plate Republics (London and Buenos Aires, edition of 1869), 9–12; Sucesión Ricardo Newton, AGN, Sucesiones No. 7217.

1870 Seguí, op. cit., 67; Sucesión J. M. Biaus, AGN, Sucesiones No. 4026.

1871 MacDonnell to Granville, Commercial No. 79, Buenos Aires, 15 July 1871, F.O. 6/304.

1872 BRPM, 22. Nov. 1872.

1875 Mulhall, op. cit. (edition of 1875), 421; Seguí, op. cit., 67.

1876 Richard Napp, The Argentine Republic (Buenos Aires, 1876), 307–309; Frederick Woodgate, Sheep and Cattle Farming in Buenos Aires (London, 1876), 23.

1879 The South American Journal (SAJ), 23 May 1879, 5.

1880 Seguí, op. cit., 67.

1884 SAJ, 4 Oct. 1884; E. R. Coni, Die Provinz Buenos Aires (Zurich, 1884).

1885 Mulhall, op. cit. (edition of 1885), 270; Seguí, op. cit., 67.

1886 G. Cormani, Argentina: Guida per l'emigrazione (Milan, 1888), 131–133.

1888 E. Zeballos, Descripción amena de la República Argentina (3 vols., Buenos Aires, 1881/1888), vol. 3, chap. 12; Censo agrícolo-pecuario de la Pcia. de Buenos Aires, 1888, 93–94.

1889 L. Guilaine, La république argentine (Paris, 1889), 135–140.

1890 Parliamentary Papers, Commercial Reports (d) Annual and Miscellaneous Series, vol. LXXXIV, 1890–91, 24; Mulhall, op. cit. (edition of 1892), 270.

1891 Parliamentary Papers, Commercial Reports (d) Annual and Miscellaneous Series, vol. LXXIX, 1892, 404–405.

1892 Parliamentary Papers, Commercial Reports (d) Annual and Miscellaneous Series, vol. LXII, 1893–94, 170–172; H. Gibson, The History and Present State of the Sheep-Breeding Industry in the Argentine Republic (London, 1893), 86–93.

1895 Seguí, op. cit., 67.

situation of any laborer probably depended very much on the quality and quantity of the food and dwelling given to him by his employer, and, therefore, generalizations could prove of little use to evaluate the living conditions of the workers.

Moreover, cash wages were not always paid to the workers in currency and at regular, established intervals. Frequently, the estanciero paid with *vales* that had to be exchanged for goods with a particular country merchant, with whom the employer generally had a permanent arrangement and quite often a standing credit. At times, the worker collected only part of his nominal wage, saving the rest, which was kept by the estanciero sometimes for several years, until it was required by the employee. This would suggest that sometimes wages were above subsistence level, and, as we shall see below, this was in fact the case, but only in certain limited periods and for a very particular kind of rural workers.[56]

There was, however, a widespread opinion in the decades prior to the 1880s that wages were high in Argentina, as compared to other parts of the world, and that, within the country, the pastoral activities were among the best rewarded jobs. Already in the late 1840s MacCann was saying, "Wages appear low but they are really high. . . . Shepherds and herdsmen received one hundred and fifty dollars per month, together with six pounds of *yerba* and some salt, and beef and mutton without stint."[57] As we have already mentioned in chapter 1, the early sixties would prove the boom years for sheep raising, and workers would benefit from the high demand for labor resulting from an expanding production. It is clear from Table 12 that there is an important increase in cash wages at the beginning of the decade, and many voices arose either in protest or in praise to point to that fact. "In France, it is hard to believe how prodigiously wages have risen in the Argentine Republic," the French representative in Buenos Aires told his superiors in Paris in 1865.[58] In the same year, *The Brazil and River Plate Mail* said: "Business is very brisk in Buenos Aires, but a great scarcity of hands is felt, and labor is unreasonably high,"[59] and later in the year, "wages are high, employment superabundant."[60] A British settler complained, "Wages are enormously high, and one of the great drains on the settler's purse."[61] And, in 1867, with the crisis already there, the Sociedad Rural explained:

We think it convenient to call the attention of stock-raisers regarding the disparity existing between the general expenses of a rural establishment and the benefits it produces. As currency is revaluated, the price of the staples offered drops, while labor expenses (laborer's wages) far from declining, increase considerably. Thus, ruin is inevitable.[62]

From the information provided in Table 12, it is apparent that wages kept their level during the 1870s and early 1880s, rising slowly in pesos but remaining quite stable in gold. The war with Paraguay, the epidemics of cholera and yellow fever, the internal armed conflicts, and the renewing of the campaigns against the Indians kept draining men from the potential labor force, thus keeping wages up in the long run, despite temporary setbacks experienced by the pastoral industry.

But the late 1880s were to see a significant drop in wages, as several factors combined to solve the problem of scarcity in the labor force. Immigration on an increasingly massive scale, technical innovations which allowed for the reduction of personnel in sheep raising, fencing, the conclusion of the age of wars and armed struggles, and the imposition of law and order, enforcing the rights of property throughout the country put an end—at least temporarily—to the decades-long problem of acute scarcity of hands. Furthermore, the approaching crisis was to restrain expansion, thus causing a drop in labor demand. In 1888 a Belgian observer remarked:

Wages are not really as high as they seem to be. In effect the price of labor is in direct proportion to the price of subsistence. Therefore, the high figures apparently reached by wages are really—when compared to the cost of subsistence—only equal and frequently lower than those of most European countries.[63]

A British envoy noted in 1891 that "wages were really usually much higher."[64] Gibson stated: "Labor is cheap in the Argentine Republic, particularly so at the present time, when the general tightness in commerce has tended to reduce the wage tariff," and although he went on to qualify his statement by remarking that there were still opportunities for the hard-working immigrants, it is obvious that gone were the high levels reached in the early 1860s.[65]

In any case, what did "high wages" mean for the peon or the shepherd immersed in the rural society of Buenos Aires during the second half of the nineteenth century? Life was solitary and hard for jornaleros, peons, and puesteros. Work to be done in the estancia lasted from daylight to sunset. There were very few days off, *ranchos* were less than comfortable, food abundant but monotonous, social life was scanty, and diversions were limited to a horse race or a game of cards and a drink at the pulpería. For la porary workers, things were harder still, as even family life was frequently out of their reach.

Yet they were paid good wages. And if they worked hard enough, managed to live on the food provided to them as part of their pay, and spent very little on drink, gambling, and other so-called "vices," they might save money and turn it into capital by buying some sheep, entering in sharecropping, and eventually finding their way out of their condition of dependent laborers.[66] But this possibility existed only at the individual level, and the situation varied greatly with time, personal skills, and luck. However, at the more abstract level of the labor force in general, reproduction of labor in the long run was in no way impaired by wage standards.

The consolidation of capitalism in any particular society requires the organization of a free labor market. This process in turn implies the definition and reproduction of the conditions of existence of a class of free men who have to sell their labor power in the market in order to be able to provide for their subsistence.

In Buenos Aires, these conditions were fulfilled during the second half of the nineteenth century, when the creation of a stable and reliable labor supply was the main purpose in the organization of the labor market. Long- and short-term policies contributed to this process, which was not void of contradictions. The long-term policies which led to the capitalist organization of the countryside are well-known. The appropriation of all available land by a bourgeoisie holding clearly defined property rights over it, and the incorporation of the land into productive use; the imposition of law and order throughout, along with the incorporation of large numbers of immigrants to increase the working population, and thus reduce the problem of perennial shortages in labor were salient features of the process.

These long-term policies were reinforced in the short run by concrete measures that contributed to keep the available working population under control, and to channel all the existing hands into wage labor. Several mechanisms were used to this effect. We have mentioned legal instruments, such as laws on vagos and malentretenidos, and the Rural Code. Other less formal, though not less effective, means of control were the prevention of extra earnings and the retention of wages.[67] As we have seen, the former was enforced to prevent workers from becoming independent of wage labor for their subsistence and the latter was a means of delaying cash payments which were always burdensome to the employer, but also a good way of ensuring the permanence of the worker until payment was made effective. This system appears as an opposite to indebtedness, which was also found in the province of Buenos Aires in this period. As Bauer remarked, "All these mechanisms seemed to aim at a systematic blocking of alternatives in order to obtain a constant and reliable work force."[68]

The question of the level of wages, therefore, is only one aspect of the problem of reproduction, and it has two sides to it. On the one hand, high wages may actually contribute to increasing the volume of the potential working population by attracting permanent immigrant labor as well as seasonal workers. On the other hand, if in any way these high wages are combined with easy possibilities of access to the means of production, then they can eventually hinder the process of reproduction of the labor force, by allowing workers to find other means of subsistence rather than selling their labor-power. In the period under study, those conditions were present only during the first decades of the sheep expansion era, and even then for wage laborers strictly speaking it was very difficult to become independent. In the 1840s and 1850s, for example, 500 sheep *al corte* could be bought with the equivalent of 14 months of the average wage of a shepherd, while 40 months were needed to purchase 300 hectares of the cheapest land available and a flock of 1,000 head. By the 1880s the same purchases demanded respectively 71 and 271 months of an average wage.

But wage labor was not the only type of engagement for shepherds, and we shall see how aparcería developed, and what were its main characteristics and contradictions.

Aparcería. The term *aparcería*, which could be technically translated as *sharecropping*, has been widely used in Argentina and the rest of Latin America. Yet the term may be misleading, as it has been applied to describe such differing systems as that of the *huachilleros* in Peru, the Italian wheat growers of Santa Fe, and several others. Only in their superficial forms may these systems be said to resemble one another, and therefore we must see what the term means in the case of the pastoral industry in the province of Buenos Aires during the period under consideration.[69] Basically, aparcería is a contract between a worker, who also acts as capitalist, providing his labor-power and part of the capital necessary to run the unit of production, and a landlord-capitalist, who supplies the land and the rest of the capital required for the enterprise. At the end of the period established by the contract, the worker is paid with a share of the product, while the landlord keeps the rest.

In the case of the pastoral industry in the province of Buenos Aires, however, we have to distinguish between two different types of contracts. Particularly during the first two decades of sheep expansion, it was not unusual for a worker to engage in sharecropping by providing only his labor-power for a certain period of time, and perhaps a very small amount of capital, sharing the expenses of the enterprise in that period (generally, this sum was discounted from the shepherd's share at the end of the period). The landlord, besides the land, provided the *whole* flock of sheep, and the capital necessary to pay for his share of the expenses. He also generally supplied the tools, equipment, and lodging—a hut—to be used by the shepherd. At the end of the period, the aparcero received his share of the product—which may have been a half, a third, or a fourth, according to the terms of the agreement—and which included not only a part of the wool, sheepskins, and tallow extracted, but also some of the lambs born during the period. The landlord-capitalist kept the rest.

This type of aparcería differs from the classic form described by Marx which could also be found in the province of Buenos Aires in the period under study. The worker in this case provided not only his labor-power and a small sum of capital to cover the regular expenses of the holding, but also he had to supply part of the flock, thus participating as a minor partner in the enterprise.[70]

In the first case, when the worker provided only his labor-power, receiving in return part of what had been produced during the same period, the relationship between aparcero and landlord strongly resembled that of wage laborer and capitalist. Actually, the shepherd, who had nothing to offer but his labor-power, sold it to the owner of the means of production in return for a wage, expressed as a share of the product of the enterprise run by the capital advanced by the landlord. As we shall see in the following chapter, it was also he who had control of the productive and circulation processes, subject to the central organization of the estancia. Our aparcero was equivalent to a wage laborer, and was thus seen by contemporary observers, who considered the part of the profit the shepherd received simply as the payment for the work he had performed.[71]

For the worker, however, the main difference between this system and wage labor lay in the fact that sharecropping could become a direct way of access to sheep, one of the main means of production. Therefore, in opening the possibility for him to have his own flock, it paved the way to his becoming a small independent producer. However, this process would take place only if a set of other factors were present to favor the worker, complementing his access to sheep with other conditions such as (a) relatively easy access to land, either in property or tenancy; (b) the possibility of starting a small unit of production at a profit; or (c) favorable terms in the circulation networks which would enable the aparcero to sell his share of the product at a good price. As we proceed with our study we shall see that only in the first decades of sheep expansion were some of these conditions present.

Contemporary observers, however, were very enthusiastic about the advantages of sharecropping. MacCann remarked in the late 1840s:

It was found to be of great advantage to give the shepherds an interest in the increase of the flocks, specially those at a distance from the principal establishment. These contracts were generally for four or five years. The individuals usually selected for this purpose were Irish laborers; to whom was given one-third of the increase and one-third of the wool, for which they took care of the flock and paid all expenses. A flock of one thousand five hundred sheep was generally committed to them, with sufficient land to feed double that number. . . . In this way a laboring man becomes a small proprietor in three or four years, having a good flock of sheep and a sum of money, saved from his share of wool and sheep sold during the contract, with which to commence an independent establishment.[72]

And he goes on to give a few examples of rapid profits made by sharecroppers. Hutchinson, F. Ford, and Mulhall used very similar words to describe the system in the 1860s. They also observed that a sharecropper could end up with a flock of his own after three or four years' work, when he could engage in *medieria* of the second type or even establish himself as an independent flock owner on rented land.[73]

So it appears that up to the mid-1860s this system of aparcería was very favorable to the laborer, who in few years could become his own master, engaging in tenant farming or even acquiring a plot of land of his own. Yet the favorable conditions that contributed to this process would gradually disappear. Already in the late 1860s, Latham remarked:

This phase of sheep farm management [medieria with no capital] in due course reached its climax, and . . . the interest or share given in lieu of wages soon became much more than an equivalent to a good wage. The interest given was reduced in new contracts to one-third, and then to one-fourth, and ultimately, according to the situation of the land and the quality of the sheep, to one-third and one-fourth of the increase, without wool; and the majority now pay their shepherds' wages in money.[74]

Born as a result of the acute shortage of an adequate labor force during the first decades of sheep expansion, aparcería with the favorable characteristics of the 1850s and 1860s proved to be a transitional phenomenon, as both its internal contradictions and external conditions led to changes in the relationship between estancieros and shepherds.

On one side, the share of the shepherd was reduced, so that by 1880 it was no longer possible for him to engage in medieria or tercieria if he had no capital. Moreover, the sharecropper gradually lost the independence he had earlier enjoyed in his business. Thus, although he had been used to selling his share of the wool, skins, and tallow to the estanciero, who fixed prices and mediated between the worker and the merchant, he could and did dispose of his share in the increase of the flock. By the 1880s, even this bit of autonomy was cut down and Gibson remarks, "Now the profit-sharer is not permitted to remove his share of the increase, but is paid for it at a fixed rate."[75]

Furthermore, other "external" conditions were also changing. Land prices swelled and therefore investment in land came to represent a much higher proportion of the total capital required for establishing a sheep farm—rising from 24% to 72% of that total betweeen 1850 and 1880 (see chapter 5). The opposite was true for sheep: in the 1850s the animals necessary to stock a farm represented 70% of the initial outlay of capital required to set up such a farm, while in the 1880s this proportion had dropped to 24%. It is clear enough that land, and not sheep, became the key resource needed in order to start a farm.

Finally, commercial and transportation networks became more and more centralized, and financial institutions more complex. These changes were not always to have a negative effect upon small producers, but seldom did they spell better conditions for their lot (see chapters 6 and 7).

These changes meant that for sharecroppers it became increasingly difficult to reach the status of independent producers. This first type of aparcería had developed during the first decades of sheep expansion as a means of attracting workers at a time of great scarcity of hands. At the time, it proved efficient for the estancieros and attractive for the workers, but as conditions in the labor market and in the pastoral industry changed, sharecropping lost its appeal for capitalists in the area under study, although it was still practiced in the more recently opened territories of the south.[76]

The second type of aparcería, in which the employee provided not only his labor-power and a proportional sum of the expenses of the enterprise during the period of the contract, but also furnished a part of the flock, was also frequently used in the area we are studying. In that system, the relationship between share-cropper and estanciero resembled that of the tenant and the land-lord. Unlike the tenant, who paid rent and kept the profits—sometimes reimbursing also a part of the rent, others having to give away part of the profits—the aparcero shared the product of the enterprise with the landowner, as both had invested part of the required capital, and while the former provided the labor force, the latter supplied the land. This resemblance was noted by contemporary observers such as F. Hinde, who says:

> A man with some money buys half a flock, taking care of all of it as payment of the rent of the land occupied by his sheep, paying half of the shearing expenses and receiving half of the wool and half of the lambs.[77]

Compared to sharecroppers of the first type, these aparceros enjoyed greater independence in the management of their stations and in the disposal of their shares. They were subject, however, to the general rules of the estancia that engaged them, and had to participate as workers in the annual events of the establish-ment—particularly the shearing, when all hands available were summoned to the head station—and not infrequently they had to sell their share of the wool, skins, and tallow to the estancieros.

For these sharecroppers, conditions also changed over the years and the possibility for achieving independence diminished grad-ually after the late 1860s. Thus, by 1880 Gibson notes:

> When the shepherd is thus actually owner of a portion of the flock, he has more say in the matter of the sale of wool and other produce, though he generally leaves the disposal of the fruits to his em-ployer.[78]

At the same time, the share of the worker dropped and in the 1880s it was hard to enter a contract on halves. It became more and more frequent to establish the share of the shepherd as a third or even a quarter of the original flock and of the annual produce. In these cases, the sharecropper was actually paying a higher rent than before, as he devoted more working time to the master's share of the flock (which was no longer one-half but two-thirds or three-quarters of the total) and he occupied a smaller part of the land with his own sheep (which represented only one-third or one-fourth of the total flock).

Besides these changes in the system, the overall transformations which we described in the first case also affected these share-croppers, thus further reducing their chances to achieve autonomy.

Aparcería seems to have been a widespread practice in the province of Buenos Aires during the second half of the nineteenth century, presenting different forms and conditions as capitalism consolidated in the pastoral industry. However, it is hard to assess how many laborers entered into this type of engagement, although it does seem that a majority of them were immigrants, chiefly Irish, but also Scotch and Basque. The censuses do not use a special category for this type of shepherds, so they might sometimes appear as puesteros/pastores or hacendados in the occupational classification of the 1869, 1881, and 1895 censuses, and cannot be separated from agricultural tenants and share-croppers in the 1888 agricultural census.

Family Labor. Up to this point we have considered the puestero, sharecropper, and shepherd as single persons employed by a master to perform a certain job. Contracts were made on an individual basis, and apparently there was no other person involved in the transaction but employer and employee. Yet in many cases that worker had a wife and children, and it was actually the whole family that worked in the puesto. This was also the case with tenants and farmers, where the family provided the main part of the labor force needed to perform the regular tasks of the unit of production. The only available data on family employment are provided by the agricultural census of 1888, in which 39,117 of the hands employed in agricultural establishments in the area under study were classified as belonging to the family of the owner of the establishment, while permanent wage laborers amounted to only 24,199.[79]

In the rural establishments of the first half of the nineteenth century there was little room for the family. Peons, herdsmen, and *arrieros* led a solitary life and only a few of them could keep a stable home for wife and children. For this period, few are the studies which refer to the role of women and the family in rural areas, and the available literature suggests that there was no or-ganic place for them in the economic structure of the province.[80] With the rising complexity in the organization of production and the incorporation of European immigrants that accompanied the expansion of sheep raising, the family became an essential part of the social and economic rural order. The need to attract im-migrant labor and the fact that sheep raising could be organized to employ the whole family led to a transformation of the old puesto in charge of one or two men who took care of 1,000 to 2,000 head of cattle. The new puestos, whether in the hands of sharecroppers or wage laborers, or even tenants and farmers, were often in the hands of families, who ensured not only pro-duction, but also reproduction of the labor force in its different aspects. Thus, not only the new shepherds were born from among these families but also they were trained in rural skills and not infrequently they were also taught law, order, and religion—an education their predecessors had lacked, to the dismay of estan-cieros and justices of the peace, who resorted to coercion to obtain discipline.

We find two different types of family labor in sheep raising: families of wage laborers and sharecroppers, and families of ten-ants and farmers. In the first case, the wage paid by the estanciero to the worker had to allow not only for his subsistence but also for his family, while the employer in fact received the produce of work done by almost all members of that family. Therefore, when referring to the wage paid to these shepherds, it is necessary to have in mind that in most cases, with the individual laborer a whole family of workers was being employed.

Actually, estancieros preferred to hire married shepherds. It was one more way of trying to ensure the stability of the labor supply. As Godofredo Daireaux put it:

> As the occupation of the shepherd compels him to live in one place, it is better to choose married than single men for the job. The latter frequently are overcome by boredom, and one day they end up in the *pulperia*, where they find the strength to spend all the conver-sation they had saved up during their solitary existence in their huts. Meanwhile, the flock remains unattended.[81]

The life of these families of shepherds was ruled by the routine of work, as home and puesto were one and the same, and the estancia was their most immediate environment. Men and boys—except the very young and the very old—were in charge of most operations regarding the shepherding of the flocks, although women also took part in minor jobs. The women's main tasks, however, were the housekeeping, nursing, and education of the young, and the care of fowls and orchard.[82] In the next chapter, further reference will be made to the relationship between these families of shepherds and the estancias where they lived and worked.

The second type of family labor was that used by farmers and tenants who were entrepreneurs of their own establishments. In this case, family labor replaced wage labor, itself producing the surplus necessary for the process of accumulation. I will analyze this type of labor when describing farming and tenant farming in a future chapter.

Notes

1. Tulio Halperin Donghi, "La expansión ganadera en la campaña de Buenos Aires," in T. Di Tella and Tulio Halperin Donghi, *Los fragmentos del poder* (Buenos Aires, 1969), 21–73.

2. República Argentina, *Registro estadístico de la República Argentina,* 1865, 56.

3. Provincia de Buenos Aires, *Registro estadístico del Estado de Buenos Aires,* 2° semestre 1854 (Buenos Aires, 1855), Table X. The word *peon* in Argentina refers to an unskilled laborer and does not imply servitude or bondage.

4. For an attempt to portray the life of the gaucho, see Richard Slatta, *Gauchos and the Vanishing Frontier* (Lincoln, Neb., 1983). For the pre-1810 period, several articles published in recent years on rural production, labor, and the gaucho in colonial Buenos Aires have generated a stimulating debate on the subject. See *Anuario IEHS,* 2 (1987) (Instituto de Estudios Histórico-Sociales de la Universidad Nacional del Centro de la Provincia de Buenos Aires).

5. Halperin, op. cit., and Eduardo Astesano, *Rosas: Bases del nacionalismo popular* (Buenos Aires, 1960); John Lynch, *Argentine Dictator, Juan Manuel de Rosas, 1829–1852* (Oxford, 1981).

6. Halperin Donghi, op. cit., 64; Eduardo Olivera, *Historia de la gana-*

dería, agricultura e industrias afines en la República Argentina 1515–1927
(Buenos Aires, 1928), 20–21; and William MacCann, *Two Thousand Miles'
Ride Through the Argentine Provinces, Being an Account of the Natural Products
of the Country and the Habits of the People* (2 vols., London, 1853), vol. 1,
99.

7. Wilfrid Latham, *The States of the River Plate* (London, 1868), 35–
40.

8. Herbert Gibson, *The History and Present State of the Sheep Breeding
Industry in the Argentine Republic* (Buenos Aires, 1893), 67–70.

9. República Argentina, *Primer censo de la República Argentina, 1869*
(Buenos Aires, 1872); República Argentina, *Segundo censo de la República
Argentina, 1895* (Buenos Aires, 1898); *Registro estadístico del Estado de Buenos Aires,* 2° sem. 1854 (Buenos Aires, 1855); *Censo general de la Provincia
de Buenos Aires, 1881* (Buenos Aires, 1883).

10. Provincia de Buenos Aires, *Registro estadístico del Estado de Buenos
Aires,* 1er. semestre de 1855, Nos. 5, 6 (my translation).

11. Xavier Marmier, *Buenos Aires y Montevideo en 1850* (Montevideo,
1967), 59 (my translation).

12. Halperin Donghi, op. cit., 48–49 (my translation).

13. For colonial Buenos Aires it has been argued that the existence of
alternative ways of subsistence for the potential workers was functional
to a market with such fluctuations in demand (see the debate in *Anuario
IEHS,* 2 [1987]). However, it can also be argued that the pattern of
demand was itself a result of the problems posed by the labor shortage,
which must have been taken into account by rural entrepreneurs when
devising the organization of their rural enterprises and the ways in which
they combined permanent and seasonal labor in their estancias.

14. See Hilda Sabato, "La formación del mercado de trabajo en Buenos
Aires, 1850–1880," *Desarrollo Económico,* 24, No. 96 (Jan.–March, 1985).

15. In this case I am concerned with the mechanisms set in motion
in order to create a stable labor supply, and therefore I emphasize the
view "from above." For a different approach to the same problem, see
Hilda Sabato, "Trabajar para vivir o vivir para trabajar: empleo ocasional
y escasez de mano de obra en Buenos Aires, ciudad y compaña, 1850–
1880," in Nicolás Sanchez Albornoz (ed.), *Población y mano de obra en
América Latina* (Madrid, 1985).

16. *Antecedentes y fundamentos del Código Rural* (Buenos Aires, 1864),
155 (my translation).

17. Ibid., 187 (my translation).

18. Ibid., 166 (my translation).

19. These testimonies confirm what has been already convincingly
argued by Tulio Halperin Donghi (op. cit., 63–67). The problem of in-

debtedness has been subject to great controversy. For a critical review of the debate, see Arnold Bauer, "Rural Workers in Spanish America: Problems of Peonage and Oppression," *Hispanic American Historical Review,* 59(1) (Feb., 1979), 34–63.

20. *Antecedentes,* 184 (the opinion belongs to Manuel López). (My translation.)

21. Ibid., 195 (the opinion belongs to Jose Thwaites). (My translation.)

22. Ibid., 166, 251 (my translation).

23. See Tulio Halperin Donghi, op. cit.

24. On these measures see: Ricardo Rodríguez Molas, *Historia social del gaucho* (Buenos Aires, 1968); Benito Díaz, *Juzgados de paz de campaña de la provincia de Buenos Aires (1821–1854)* (La Plata, 1959); Adrian Patroni, *Los trabajadores en la República Argentina* (Buenos Aires, 1897).

25. MacCann, op. cit., I, 145.

26. Provincia de Buenos Aires, *Código rural de la Provincia de Buenos Aires* (Buenos Aires, 1865).

27. Ibid., Article 289 (my translation).

28. This disapproval was expressed in different ways, ranging from occasional references in books or articles that referred to rural life, to complete texts dedicated to condemning the system and its consequences. Such was the case of famous literary works, like the *Martín Fierro,* and of articles of denunciation, like those published by Estrada in the *Revista Argentina.* Some of this literature is mentioned and summed up in Rodríguez Molas, op. cit.

29. *The Brazil and River Plate Mail,* 21 June 1865, 368.

30. I am not concerned here with the push factors that prompted these Europeans to emigrate. However, for a complete analysis of Irish immigration to Argentina, see Juan Carlos Korol and Hilda Sabato, *Como fue la immigración irlandesa a la Argentina* (Buenos Aires, 1981).

31. See, for example, MacCann, op. cit., I, 144–46.

32. MacCann, op. cit., I, 70–71.

33. C. Chaubet, "Buenos Ayres et les provinces argentines," *Revue Contemporaine,* 29 (1856–57), 247 (my translation).

34. A. Hournon, "Le Río de la Plata—situation présente," *Le Correspondant,* 48 (1859), 705 (my translation).

35. *Parliamentary Papers, Commercial Reports* (b) Embassy and Legation, vol. LXIX, 1867, 319–20.

36. B. Mitre, "La inmigración espontánea a la República Argentina," in *Discursos del General Mitre* (Buenos Aires, 1870), 46 (my translation).

37. *Parliamentary Papers, Commercial Reports,* (b) Embassy and Legation, vol. LXXXIX, 1881, 159.

38. *Parliamentary Papers, Commercial Reports,* (d) Annual and Miscellaneous series, vol. XCII, 1893–94, 170–71.

39. See, for example: M. Iriart, *Corsarios y colonizadores vascos* (Buenos Aires, 1945); M. G. and E. T. Mulhall, *Handbook of the River Plate Republics* (Buenos Aires, 1875) (information on the province of Buenos Aires); Archives des Affaires Etrangères (France), *Correspondance commerciale,* Buenos Aires (consular), vol. 4, 1855–58, 95–99, 30 July 1855; *Parliamentary Papers, Commercial Reports* (b) Embassy and Legation, vol. LXIX, 1867, 319–25, and vol. LXXXIX, 1881, 157–61; *Globus,* 11 (1867), 252; Comte de Gobineau, "L'émigration européen dans les deux Amériques," *Le Correspondant,* 89 (1872), 208–42; Emile Daireaux, *Vida y costumbres en el Plata,* (2 vols., Buenos Aires, 1888), vol. II, 5.

40. On Irish immigration and sheep farming, see Juan Carlos Korol and Hilda Sabato, op. cit.

41. On government action regarding immigration, see, for example, Miguel Bejarano, "Inmigración y estructuras tradicionales en Buenos Aires (1854–1930)," in T. Di Tella and T. Halperin Donghi, *Los fragmentos del poder,* 77–149; Gustavo Beyhaut et al., "Los inmigrantes en el sistema ocupacional argentino," in T. Di Tella et al., *Argentina, sociedad de masas* (Buenos Aires, 1965), 85–123; Gastón Gori, *La Pampa sin gaucho* (Buenos Aires, 1952); James Scobie, *Revolución en las pampas: Historia social del trigo argentino* (Buenos Aires, 1968).

42. There are many such reports. See, for example, Archives de Affaires Etrangères (France), *Correspondance commerciale,* Buenos Aires (consular), vol. 7, 1865–67, 333–34, 12 Feb. 1867, and 451–52, 12 August 1867; G. Gaudelier, *La vérité sur l'émigration des travailleurs et des capitaux belges* (Bruxelles, 1889), 38; *Scientific American,* 38 (n.s. 6) (April, 1878), 213. On this subject it is interesting to point out that in 1867 the British Emigration Board recommended British subjects to emigrate to British colonies rather than to a foreign country. Colonial Office to Foreign Office, London, 26 March 1867, F.O. 6/271.

43. *Código rural,* Article 222 (my translation).

44. Ibid., Articles 223–37.

45. *Censo agrícolo-pecuario de la Provincia de Buenos Aires, 1888.* According to the census, for the whole province 180,652 permanent laborers (74,811 wage laborers with board, 6,739 without board, and 99,102 belonging to the family of the producer) and 219,500 temporary workers were employed in the agricultural sector.

46. Although shearing is a highly skilled job, shearers as a category are not included in the national or provincial censuses. The 1881 census was taken at the beginning of the shearing season, but we find no trace of the job in the categories included, which were based on the French

classification. This makes very difficult any attempt at quantification, as shearers were then included in the very broad category of jornaleros.

47. Gibson, op. cit., 71–75.

48. See Benjamín Vicuña Mackenna, *La Argentina en el año 1855* (Buenos Aires, 1936), 123; Latham, op. cit., 32; N. MacLeod, "The Life of a Sheep-Farmer in the Argentine Republic," *Good Words*, 12 (1871), 712–22; Emile Daireaux, op. cit., II, 314–15; Godofredo Daireaux, *La cría de ganado en la estancia moderna* (Buenos Aires, 1908), pp. 224–41; José Hernández, *Instrucción del estanciero* (2nd ed., Buenos Aires, 1964), 256–59.

49. Gibson, op. cit., 93.

50. The province of Buenos Aires has been traditionally an area of attraction of internal migrants. During the nineteenth century, the transformation of the local economies of the interior and the expansion of production in Buenos Aires led to a constant flow of laborers toward the province, particularly from the depressed areas of Santiago del Estero, but also from the neighboring provinces of Santa Fe and Córdoba. By 1869, provincianos constituted around 10% of the native population and 8% of the total population of the province of Buenos Aires, excluding the capital. In our thirty counties north of the Salado River, the figures were 8% and 7% respectively. In 1881, they represented 6% of the native and 5% of the total population of the province, while north of the Salado the proportions were also 6% and 5% respectively. However, as none of these censuses were taken during the period of peak labor demand, probably these figures do not include seasonal migrants, who came to Buenos Aires for the shearing season and later returned to their native provinces or spent the slack months in the capital as urban laborers (*Primer censo de la República Argentina, 1869*, and *Censo general de la Provincia de Buenos Aires, 1881*).

51. Gibson, op. cit., 86–87.

52. Latham, op. cit., 26–28.

53. Emile Daireaux, op. cit., 310–11; Estanislao Zeballos, *Descripción amena de la República Argentina* (3 vols., Buenos Aires, 1881), vol. 3, 303–20; Gibson, op. cit., 67–71.

54. There are quite a few examples in Zeballos, op. cit., vol. 3. See also chapter 4.

55. L. Dillon, *Twelve Months Tour in Brazil and the River Plate with Notes on Sheep-farming* (Manchester, 1867); also Zeballos, op. cit., vol. 3, chap. 12.

56. On the payment with vales, see Emilio Delpech, *Una vida en la gran Argentina* (Buenos Aires, 1944), 48; Thomas Hutchinson, *Buenos Aires y otras provincias argentinas* (Buenos Aires, 1945), 313; Etienne de Rancourt, *Fazendas et estancias: Notes de voyage sur la République Argentine*

(Paris, 1901), 265–66. On workers' savings, see Korol and Sabato, op. cit.

57. MacCann, op. cit., 24.

58. Archives des Affaires Etrangères, *Correspondance commerciale*, Buenos Aires (consular), vol. 7, 1865–67, 16–17, 12 March 1865 (my translation).

59. *The Brazil and River Plate Mail*, 23 May 1865, 318.

60. Ibid., 6 July 1865, 392.

61. R. Seymour, *Un poblador de las pampas* (Buenos Aires, 1947), 39.

62. *Anales de la Sociedad Rural Argentina*, 1 (1867), 75 (my translation).

63. Caudelier, op. cit., 38 (my translation).

64. *Parliamentary Papers, Commercial Reports* (d) Annual and Miscellaneous Series, vol. LXXXIX, 1892, 404.

65. Gibson, op. cit., 96. No comparison with urban wages is possible as there are no systematic studies on the subject for the years before 1882.

66. *Parliamentary Papers, Commercial Reports* (d) Annual and Miscellaneous Series, vol. LXXXIX, 1892, 404.

67. Regarding the prevention of extra earnings, see for example Juan M. de Rosas, *Instrucciones a los mayordomos de estancias* (Buenos Aires, 1942), and the comments of Halperin on that text (Halperin, op. cit., 62–63). Examples of retention of wages appear in almost all accounts of estancias found in the *Sucesiones*. See particularly those of Ricardo Newton (1868), Guillermo White (1866), and Juan Acebal (1878), AGN, Sucesiones No. 8760, 7217, and 3695 respectively. A concrete example of indebtedness is found in Archivo Senillosa, AGN, Sala 7, 2-6-13, "Estado de gastos q' se an echo en Establecimiento del Arroyo Chico del Sr. Don Felipe Senillosa, 1 Marzo y 1 Junio 1842." For a discussion on indebtedness, see Arnold Bauer, op. cit., 48.

68. Bauer, op. cit., 48.

69. In this chapter, *aparcería* will be analyzed from the point of view of the worker employed under such contract, while in the next chapter I shall consider its implications for the estancieros.

70. See Karl Marx, *El Capital* (Mexico City, 1946), III, chap. 47.

71. See MacCann, op. cit., I, 282; Hutchinson, op. cit., 313; Gibson, op. cit., 86.

72. MacCann, op. cit., I, 281.

73. *The Brazil and River Plate Mail*, 21 June 1864, 322 (article by Hutchinson); *Parliamentary Papers, Commercial Reports*, (b) Embassy and Legation, vol. XLIX, 1867, 333; M. G. and E. T. Mulhall, *Handbook of the River Plate Republics* (Buenos Aires, 1869), 18.

74. Latham, op. cit., 25.

75. Gibson, op. cit., 274.

76. Godofredo Daireaux, op. cit., I, 282–83.

77. Cited by Hutchinson, op. cit., 313.

78. Gibson, op. cit., 92–93.

79. *Censo agrícolo-pecuario de la Provincia de Buenos Aires, 1888*, 55–92.

80. See, for example, John Lynch, op. cit., 103–4.

81. Godofredo Daireaux, op. cit., 132 (my translation).

82. There is very little reference to the role of women and the family in studies that deal with different aspects of the rural society in nineteenth-century Buenos Aires. Actually, the only specific work devoted to this subject is *La mujer en la pampa*, by Maria Teresa Villafañe Casal (La Plata, 1958), which presents a rather romantic overview of the different roles played by women in the rural areas. A description of the life of families of Irish shepherds is found in Korol and Sabato, op. cit., and more general references are found in accounts of contemporary observers, such as MacCann, op. cit., I, 24–25; Zeballos, op. cit., vol. 3, 303–20; Mulhall, op. cit. (1869), 161–62.

4

Estancias and Estancieros

The inconveniences attending estancia life are many, but
we know of no legitimate enterprise now-a-days which
offers a surer yield or a safer investment than that of sheep
when properly attended to.
—*The Brazil and River Plate Mail*, 22 November 1865.

In pre-1850 Buenos Aires province, estancias had become the rural enterprise par excellence. Land and cattle were the key resources upon which post-Independence wealth was built. A mercantile class, which at the rise of the nineteenth century was more interested in commercial pursuits than in productive activities related to the rural areas, had nevertheless sought in the late colonial era the benefits of cattle raising, which provided hides and salted meat for export first to complement and later to replace the staple export of colonial Buenos Aires—Potosí bullion. The hinterland of Buenos Aires became the key area for this expansion which reached its more dynamic phase during the Rosas regime. Large tracts of land were appropriated and put into productive use by the simple device of building a hut or rancho, putting a few men in charge of several thousand head of cattle, and claiming the rights of property over land and stock. Thus, the first estancias were born.

By mid-century, estancia management had acquired a higher degree of complexity, but always the guiding formula was abundant land, large herds of semiwild cattle, a few working hands, and a low investment in physical assets.[1]

At first the introduction of sheep seemed to bring about little change in the overall organization of rural production. In the 1820s and 1830s very few estancieros considered sheep raising as an important item in their rural activities. More often than not, sheep of very low quality (*criolla*) were kept simply as a complement to cattle, and were seldom shorn.[2] Only a small group of enterprising men, most of them immigrants seeking a profitable field for investing their capital which had been made in other activities, saw a possibility in the expanding international demand for wool, and started to promote the development of sheep breeding and wool production in the River Plate area.[3] Based on the pattern of the "old estancia," a new type of enterprise started to take shape, and in a few years it had spread over all the province, particularly north of the Salado River. A closer look at these establishments would reveal a motley picture. Units of production devoted to sheep raising could be large estates of 30,000 hectares or small farms of 500 hectares or anything in between. They could be run by the owner of the land or by a tenant; they could employ wage labor or rely mainly on domestic labor; and they could combine all these features in different patterns, with each establishment unique.

How can we systematize this variety of cases and still avoid easy simplifications? Bearing in mind that one of the main purposes of this work is to analyze the process of capital accumulation in the period and area under study, pastoral concerns can be included in two main types of enterprise—the estancia and the sheep farm.[4]

Estancias were capitalist enterprises that produced for the market by employing wage labor, but also combining it with other forms of labor (i.e., different forms of aparcería). The main objective of the estancieros was maximization of profits in order to ensure the process of accumulation. Although farms also produced for the market, they were mainly family concerns which relied almost exclusively on domestic labor, and made only occasional use of wage labor. Farmers were not peasants, however. They were not only concerned with the survival of their unit of production and the subsistence of themselves and their families, but also with the reproduction and expansion of their enterprise.

This categorization assumes that estancieros and farmers owned the lands upon which their enterprises were organized. Therefore, for each individual enterprise, returns included not only profits but also rent and differential rent.[5] Moreover, the increase in the value of land became an important item in the process of accumulation. Purely theoretical considerations would require that, for analytical purposes, landlords and capitalists should remain as separate social actors. But I have deliberately chosen a classification that will allow me to portray the type of establishment, more widespread in our area, in which ownership of land and pastoral enterprise became one single business, and were considered as such by contemporaries. However, establishments run on rented land will not be left out of our description, as *arrendamiento* was an extended practice in the period and area under study. Tenant estancieros and tenant farmers will be considered here as particular cases of the more general categories of estanciero and farmer.

In the following pages I shall discuss the main characteristics of the estancia as a unit of production. The next chapter will be devoted to the analysis of sheep farms.

The Estancia

Estancias and estancieros never fail to appear in any description or interpretation of Argentine society. The literature on the subject has put forward many different hypotheses on the role of estancieros throughout their history, but their relevance as a class has never been questioned. Whether described as a landlord class of feudal or semifeudal behavior or included in a bourgeoisie driven by capitalist objectives, estancieros are always considered among the powerful of the country. I shall not deal here with the more general discussion on the nature and role of estancieros in Argentine history, but with a more limited problem—that of the organization of estancias in the era of pastoral expansion and the characteristics of sheep estancieros in the period and area under study.[6]

Since the eighteenth century, the word *estancia* has been applied in Argentina to many different types of rural enterprise. In the province of Buenos Aires it has been generally used to refer to capitalist rural concerns, whose organization and main productive activities varied with time—from the primitive cattle-raising and hide-producing estancias of the late colonial era to the present highly mechanized grain- and cattle-producing establishments. And although some estancias actually went through all the transformations leading from one to the other type of establishment, many more probably survived through only one or two stages in this development.

Therefore *estancia* as a synchronic concept has different connotations in the different periods and areas where it is used. In this sense, for example, the concept of *estancia* in the eighteenth century has little to do with the term *estancia* used to refer to present-day estancias in the province of Buenos Aires. However, the concept has a diachronic dimension, and in considering it, we shall find that the word *estancia* is used today in a way that contains and includes the preceding meanings of estancia.

In this text, I shall try to portray the main features of the estancia in the era of pastoral expansion, and therefore, I shall talk about sheep estancias in order to depict those characteristics which pertained to establishments primarily devoted to pastoral activities. Throughout the period and area under study, however, estancias evolved from a first stage in which sheep were introduced as complementary to cattle, to a second stage in which sheep were decidedly predominant, and then reached a final stage in which pastoral activities gave way to grain and cattle production. Although the pace of development could vary for each estancia— and probably many of them did not go through all these stages— we must bear in mind the main direction followed by this process of transformation. Therefore, the word *estancia* in our text must be considered in its diachronic dimension.

Our analysis will center on pastoral establishments mainly dedicated to wool production, and we will make only side references to cross-breeding cabañas or to estancias where cattle played a relevant role. As will be seen below, wool production and sheep raising became a single business for these establishments, as reproduction of stock played an important part in the process of capital accumulation.

Estancias varied in size as well as in cattle-carrying capacity, as not all the soil of the area was homogeneous nor was water evenly distributed throughout. Nevertheless, we can safely assert that most of the sheep were raised in estancias of over 5,000 hectares, but that in numbers, establishments under that size were predominant.[7]

In the counties under study, the carrying capacity of land was estimated at an average of 3 to 4 sheep per hectare, although this varied not only with the natural conditions of the environment and soil but also with the improvements incorporated by the estanciero, such as wire fencing or water supplying devices. Therefore, the number of sheep in these establishments could vary greatly, and, indeed, it did, for we have found estancias of under 5,000 head while others could raise over 100,000. Besides sheep, it was necessary to have horses and mares, which were indispensable in field work. Their number, of course, varied according to the size of the establishment, ranging from under 100 to several thousand. However, these variations in the scale of production had little influence on the way the process of production was carried out and only in very extreme cases did a qualitative difference arise.

The Organization of the Enterprise

Whether large or small, most estancias were organized in a similar way.

> An establishment of this class (estancia) consists of an estancia house of more or less pretensions in size, round about which may be seen generally plantations of paraiso, acacia, poplar, willow, mulberry, some few sturdy "ombu" trees, and in some cases peach-plantations, out-offices, sheds or "galpones," horse "corrales" or pens, sheep corrales. The "poblaciones" . . . stand about clear in the plain and are seen for considerable distances. . . . On different parts of an estancia there are erected the huts of the shepherds.[8]

Such was the basic aspect of sheep estancias throughout the period. The pattern allowed for variations, however, and in the early 1850s most establishments were content to have a two- or three-room central house made of adobe, and adjacent to it a precarious *ramada* which served as a shed for different purposes, maybe one or two folds to harbor the main flocks under the care of a peon, a kitchen for the workers, a well for water, and a few trees. On the periphery of the property, the one-room mud huts of the shepherds could be found, with little else around them but a fold, a couple of trees, and a well. A ditch of varied length and importance sometimes surrounded certain vital areas of the estancia. Three decades later, only a few establishments kept such a precarious look, and most inventories of the 1880s included a well-built brick house with from five to perhaps over twenty rooms for the owner and his family; another brick house for the manager and his family; a house for the workers; two or more wooden or brick sheds (one for the shearing, the rest for storage purposes); a well-furnished kitchen; an orchard where trees and vegetables were grown; several folds for the different flocks and types of sheep, and, finally, in the periphery the shepherds had their mud huts replaced by modest brick two-roomed houses. The introduction of wire fencing further changed the aspect of these estancias, and even a rearrangement of the huts was accomplished in order to take full advantage of the benefits that this innovation made possible.[9]

Thus, an establishment belonging to the same family could have had in the 1880s very little resemblance to its predecessor in the first half of the century. Such was the case with the estate of the Girados in Chascomús. The inventory of their property in 1822, 1858, and 1881 is an excellent example of this transformation.[10]

Francisco Girado's estancia was a modest establishment where approximately 1,000 head of cattle, 600 sheep, and three herds of horses and mares were raised in the early 1820s. A large hut with "a window and a door" (sic) for the owner, a smaller one for the overseer, a well, and a kitchen with two ramadas were all that was to be seen in the way of buildings. A few trees and an old ditch completed the picture. Within the house an old boiler, one large and two small iron pots, a kettle, two roasting sticks, a tub, a pail, five plates, eleven forks, nine spoons, two mortars, a grinding stone, two bags of salt, a hoe, a spade, an adze, a bed, two tables, a bridle, spurs and a poncho. Also there were a couple of carts which were used for transportation.

We find the family's estate greatly enlarged in the late 1850s, with one of its members, Elias, owning approximately 6,500 hectares. The dwelling house was a three-room edifice well provided with furniture in bedroom, living room and dining room. Two kitchens with adjacent ramadas; an outdoor oven; the thatched hut of the overseer; a shed with a complete set of tools including five spades, four axes, a sickle and the like, and also including a scale and two dozen pairs of shearing scissors; folds and palings completed the outbuildings of the central area of the estancia.

Next to it, the orchard had a large number of trees (16 montes) and included a hut, kitchen and well for the worker in charge. There were six puestos in the periphery of the property—all consisting of a mud hut with thatched roof, a sheepfold, a ditch, and a variable number of trees. Around 2,000 meters of ditch and fence had been built, along with a wall to guard the house from the periodical floods of the Laguna de Chascomús.

The sons of Elias—Federico and Ceferino—inherited the estate. When Federico died in 1881, their estancia had gained 1,000 hectares, and a number of very significant changes had taken place. The whole property was wire-fenced, and within it several of the sheepfolds were also protected by wire. The main buildings now included a flat-roofed dwelling house of brick with seventeen rooms, vestibule, carriage house, carpenter's shop, kitchen, water closet, well, garden, and courtyard; kitchen and room for the manager; kitchen for peons and room for the overseers; two sheds; a second well; a pigeon house; a bread oven; a brick shed for purebreds; another shed with thatched roof for well-bred cows; several sheepfolds; a special enclosure for a purebred horse, and another for ostriches ("avestruces del Africa"). All these buildings were furnished with considerable comfort, and tools of various sorts were available in the sheds. Among other utensils, the following machines are listed: washing machine, mowing machines, machine to kill *vizcachas*, another machine to fence, and yet others to thresh maize and to hatch eggs. By the way of vehicles, the inventory listed eight different types of carts and wagons. The estancia had twenty puestos, all provided with a two-room brick house, with tin roof, kitchen and ramada, a well, a *quinta*, and a sheepfold.

Changes were not limited to the physical equipment, for the livestock of the 1880s showed considerable improvement also. In the 1820s, Francisco Girado had raised 600 sheep and 1,000 head of cattle of very low quality, but in the 1850s, Elias' flocks consisted of almost 10,000 mestizo sheep, each valued at between 0.6 and 1 golden peso, and his herds of over 6,000 head of cattle were valued at 38 golden pesos per head. The Girados were gradually shifting to sheep raising, but they always kept an interest in cattle. Federico's establishment in 1882 listed over 2,000 head of cattle, including some purebreds, over 28,000 mestizo sheep, valued at

1.2 golden pesos each, 2,500 fine crossbreds valued between 2.8 and 6 golden pesos, and twenty rams considered to cost 60 golden pesos per head.

In spite of these significant changes introduced through the years, the organization of pastoral establishments remained basically the same until the introduction of the paddock system at the end of the century. Sheep were divided in flocks of 1,500 to 3,000 and each flock was placed under the charge of a shepherd.

> The huts of the shepherds, called *puestos,* are located in the highest spots at the borders of the estate. The distance between them is around one kilometer, so that sheep may find an ample space in front of them to wander without mixing with other flocks. Therefore, each puesto has around 200 hectares.[11]

Not infrequently these flocks consisted of mestizo sheep of all ages, and few were the estancieros who separated them according to quality, age and state of the animals. Most estancieros, however, did include a score or two of rams with each flock in order to ensure reproduction. Furthermore, each puesto had a few horses (not always the property of the estanciero) that the shepherd used in his job.

The daily work of these shepherds has been described in chapter 3, so let us only add a few reminders here:

> His duties are to tend his flock day and night, to keep it from mixing with other flocks with runs on the same estate when there is no divisionary fence between the several runs; to keep it free from scab and other contagious diseases; to keep dogs off, and see that no sheep wander astray; in short, to generally shepherd his charges. All this he does upon horseback, and dogs are seldom employed, as they run wild and cause great havock.[12]

These jobs would be reduced with the introduction of the paddock system, but even in the late 1880s the old methods were still widely used in the countryside.

Besides the livestock in the charge of these puesteros, generally the head station kept one flock under its direct control. More often than not, this flock was the best sheep that were kept, and when the estancia was well enough provided to include some purebreds among its livestock, they also were specially taken care of in sheds and pens close to the head station. When breeding became the main purpose of the establishment, it was known as a cabaña, and in such a case, the organization of the head station was more complex than in the rest of the estancias, and sometimes a few of the puestos could also be put in charge of entire flocks of fine-bred sheep and rams.

The main events in the working life of an estancia were the shearing, which in our area took place from October to December, and lambing, twice a year. Other annual jobs included the marking of lambs and new sheep; curing of the scab and other diseases; maintenance of buildings, fences and corrals, and so on. After the shearing, which we described in chapter 3, we will quote Gibson on further matters:

> The breeder has two important matters to which to attend. The first of these is the scrupulous revision of his stock, to cure the scab and any other malady, and get all such disorders well in hand while the wool is still short, the stock healthy and strong, and the ewes empty. He should particularly endeavor to get his great enemy, the scab, subjugated, and daily revision and periodical dipping are matters of first necessity.[13]

The way in which the curing of the scab was done depended very much on the equipment of the estancia, but the recommended method was to construct a bath wherein to pour the remedial solution and dip the flocks within twenty days after shearing.[14]

The selection of the stock to be sold in the market also had to be done in the first months of the year. Lambing took place both in April–May and July–August, and earmarking and castration was done four to six weeks after lambing. "During the winter season the sheep breeder finds time to attend to the upkeep of this establishment, the repair of the buildings and fences, ditching, attending to the tree plantations and similar occupations."[15]

These jobs were performed in two different stages: everyday routine tasks as well as lambing and sometimes curing were done by the shepherds in the area of their puestos, but the shearing required all the available hands plus the temporary ones to work around the head station. This area was also the usual working place for peons and servants attached to the main buildings.

Both permanent and temporary laborers were employed by the estancia to perform the tasks that have been described. We have referred to their different roles and the types of engagement they entered into. Here I shall attempt to advance on such considerations by analyzing the way in which these forms combined in every unit of production and how the process of production was organized.

Management and Labor

At the apex of the internal organization was the owner of the establishment who had overall control of the process of production. His actual participation in the process varied greatly in each case, but his concern regarding the development of the enterprise seems to have been deeper than the absenteeism of many estancieros would seem to suggest. Actually, most estancia owners lived in the countryside, either on their estates or in nearby country towns, thus exerting personal control over their business. Only top estancieros needed to fix their permanent residence in Buenos Aires in order to attend to their multiple business interests.[16] Apparently very few of them left their rural concerns entirely in the hands of their managers. Daireaux explains:

> Large landowners are not indifferent stock raisers who at the end of the year are satisfied to receive the produce of their estate from the hands of a manager. On the contrary, frequently they control their stock and day by day they know what is going on in their establishment. . . . From their office in the city—which they attend every day—they deal with all matters pertaining to the administration of their fortune. In that same office they sell hides, wool and stock, they operate with capital, and they make a commercial activity out of the business of managing a large fortune.[17]

Most estancieros kept the key decisions regarding their investments, sales, and management in their own hands, traveling frequently to their estancias, and maintaining an abundant correspondence with their mayordomos. Their concern regarding their rural enterprises sometimes went far enough to induce them to travel abroad in order to learn new techniques, to read about sheep raising in other parts of the world, and even to write their experiences and suggestions either in books—like those of Gibson, Hernández and Zeballos—or in articles that were published in newspapers and journals. The *Anales de la Sociedad Rural* are full of such articles written by well-known estancieros like Olivera, Senillosa, Jurado, and several others.[18]

A well-chosen manager was the first requisite to be filled by an estanciero if he wanted a successful enterprise. The manager was the personal representative of the estanciero while the latter was away, and had control over the different aspects of the process of production.

> An estancia . . . must be subordinated to one single hand, in order that its administration be achieved according to the desires of its owner. This director, this manager of the estancia is the *mayordomo*. He is in charge of running the establishment. His immediate subordinate, the *capataz*, is in charge of leading the peons to execute the orders he receives.[19]

The extent of the manager's responsibility varied greatly in each case, as it depended entirely on his relationship with the owner. In most cases it was the manager who decided on his own in such things as hiring of extra hands, minor repairs on buildings and equipment, buying of food and other supplies; but he decided in consultation with the owner, upon more important matters, such as time of curing and shearing, hiring of permanent laborers, replacement of stock or equipment, and sometimes sales of produce.

Except in unusual cases when a bookkeeper was employed, it was the mayordomo who kept day-to-day accounts, and who periodically sent them to the owner. He was in charge of all cash transactions in the estancia, and the necessary money for this was given to him either personally by the owner or by someone of the owner's family, sent through the bank or drawn from a local merchant where the estancia had standing credit.

Not all accounts were equally well kept, however. It is possible to find examples of extreme care in this regard, like that of M. Farrel, manager of Richard Newton's establishment La Posta de Vázquez, who sent periodical accounts to his employer, separating regular expenses and income in two columns, month by month, and keeping a separate record for all transactions with shepherds and sharecroppers.[20] Not always, however, were managers so careful, and in the correspondence between Enrique Gubba, manager of the Estancia del Venado, and his master, Don Felipe Senillosa, Sr., we have an example of the more chaotic way of keeping accounts in most establishments. Gubba used to mention expenses and income in his frequent though irregular letters to Senillosa, but made no attempt to organize his accounts more systematically.[21]

Managers were in charge of handling all matters concerning the rest of the personnel, which comprised one or more overseers, a variable number of peons and servants attached to the head station, the shepherds in charge of puestos, and the seasonal workers. Some establishments also included a second manager, a bookkeeper, a shopkeeper if there was a store within the estancia, and one or more specialized workers, like quinteros, carpenters, and butchers.[22]

The regular tasks of the sheep estancia were carried out by peons and puesteros controlled by the overseers. Each shepherd was in charge of a flock and of the upkeeping of his puesto, but he could also perform extra jobs, and he was asked to do so and generally paid extra, during times of high demand of labor like the shearing season. Peons could be hired to attend to specialized jobs—galponero, quintero—or to be ready to perform any task indicated by the overseers. Extra hands were generally employed for specific purposes, such as shearing, repairing a ditch, digging a well.

In chapter 3 we noted that although the more widespread type of engagement for workers was that of wage labor, there were also other forms which appeared under the wide denomination of aparcería. Actually, most estancias combined wage labor—both permanent and temporary—with sharecropping in its different forms: the first to remunerate peons, servants, and seasonal workers; the second mainly to reward shepherds in charge of puestos.[23] Examples abound of the combined use of different types of engagements in sheep estancias.

As early as 1847, we find Peter Sheridan's heirs disposing of the stock of Los Sajones as follows:[24]

On shares to Mr. Lovell, he paying ¼ of expenses
and receiving ¼ amount of produce and increase.....15,206 head

To Mr. Crampton, he paying all his own expenses
and receiving ¹/₄ amount of produce and increase..... 6,150   "
To Mr. Moore, he paying ¹/₃ of expenses and
receiving ¹/₃ amount of produce and increase.......... 1,871   "
To Martins to take care of on condition of getting
permission to keep on the land a similar number
of his own   990   "

In this particular case, the contracts were signed on a five-year basis, but apparently the deal turned out to be a failure, for all four men left the estancia before the third year was over. In this case, Lovell himself had subcontracted four *medieros* who bought half of the stock, shared expenses (one half) and obtained half of the produce and increase.

In the 1860s we may mention, among others, Henry Bell's estancia La Adela in Chascomús, where in 1862 the 13 puestos were distributed as follows: 10 on thirds to workers without capital, 2 on halves to workers with capital, and 1 to a shepherd on wages;[25] Felipe Girado's estancia Paso del Villar, also in Chascomús, where in 1863 of 20 puestos, 13 were in the hands of workers who were engaged as *medianeros* with capital, 3 in those of *medianeros* without capital, and 4 in those of shepherds on wages;[26] Samuel Wheeler's establishment in Navarro, where, besides a peon, a cook, a manager, and a quintero paid in wages, 5 medianeros with capital were in charge of the corresponding puestos.[27] Newton's accounts of his establishments La Posta de Vázquez, Vista Alegre and Santa María also show us how in the late 1860s he combined wage labor with aparcería in his estancias.

In the 1870s, Acebal's La Segunda included 34 puestos, 7 in charge of medianeros with capital, 12 of *tercianeros* with capital, 1 a tercianero without capital, 10 given away to workers on fourths, without capital, and 4 to shepherds on wages.[28]

In 1882, Federico Gandara's estancia Vitel had 41 puestos, in charge of 15 shepherds on thirds without capital, 9 on thirds with capital, and 17 medianeros with capital,[29] and in 1886, in Mercedes, Agustín Zemborain's establishment was organized into 8 puestos: 3 in the hands of shepherds on fourths without capital, 1 on thirds with capital, and the rest of puesteros on wages.

In the previous chapter we analyzed the transformation of the

Table 13
Estimated Annual Income of Aparceros[a] and Wage Laborers[b]
(1845–1885) in the Province of Buenos Aires
(in Golden Pesos)

Years	Aparceros' (1st type) cash income[c]			Aparceros' cash income plus estimated value of share of stock received			Wage laborers' cash income
	$1/2$	$1/3$	$1/4$	$1/2$	$1/3$	$1/4$	
1845–1854	138	92	69	264	176	132	98
1855–1864	281	187	140	437	291	220	126
1865–1874	293	195	147	559	372	280	160
1875–1884	394	263	197	670	447	335	188

[a]Calculations are based on aparceros in charge of a flock of 1,500 to 2,000 sheep.
[b]Average income for each period calculated on the basis of annual wages as per Table 12, chapter 3.
[c]Corresponds to their share of produce, excluding stock.

different forms of aparcería throughout the period under study; we shall now refer to the advantages and disadvantages they had when compared to wage labor, from the point of view of capital accumulation at the level of the individual enterprise.

Since the very early days of pastoral expansion, a shortage of skilled labor was one of the main problems encountered by sheep estancieros in organizing their establishments. Aparcería of the first type (see chapter 3) was most attractive for the laborers, as it proved more rewarding than wage labor (see chapter 3 and Table 13). Therefore, it was offered by the estanciero to recruit the right kind of workers. We have already seen that this method proved quite successful, as a goodly number of Irish and Basque settlers entered into this type of arrangement, thus providing the estancias with industrious families of shepherds. Also, Argentine workers soon started to flow to these jobs as well.

One of the main advantages that contemporaries saw in aparcería was that it almost ensured full dedication on the part of the worker to his job, as his reward depended entirely on how much he could produce in a year. In establishments of such an extension as most sheep estancias of Buenos Aires, where shepherds were stationed far from the main buildings and for most of the time out of sight of the manager and overseers, to make remuneration depend on productivity seemed reasonable enough. Furthermore, this method implied a low outflow of capital throughout the year, as except for the advances the shepherd might require in order to cover the necessaries of life, most of the reward was either paid once the produce was sold, or never cashed, as in the case of the share of stock kept by the sharecroppers as part of the deal.

In the face of these advantages, two questions deserve consideration, however: (a) the cost of aparcería as compared to wage labor, and (b) the excessive independence of the sharecropper in the process of production and circulation of his staples vis-à-vis the central organization.

One obvious advantage of aparcería was the variability of its cost. Contrary to wage labor, which appears as a fixed cost during a certain period of time, its price regulated by supply and demand of labor at the beginning of the period, sharecropping had a cost which was determined at the end of the year by the price of the produce, so that the cost of labor varied according to the price obtained by the estanciero for his staples.

In the first decades of pastoral expansion, the low prices of stock and wool kept a check on the share received by sharecroppers, and when those prices started to climb in the late 1860s, the income of aparceros was lowered by the simple device of diminishing their share. We have already said that medianeros without capital were found only in the early days of the business, but in the 1880s most sharecroppers received only a fourth of the produce.

Nevertheless, except in years of crisis with very low prices of wool, high mortality of flocks, etc.—aparcería meant higher costs than wage labor. We have constructed Table 13 to show the average annual income received by workers who were under different contracts of aparcería, as compared to average wage of shepherds. The data on this table follow from the calculation on income and costs of production (included in Tables 15 and 16) and are only an approximation of what actual figures might have been in the different decades. Fluctuations caused by year-to-year variations in the international price of wool and other byproducts do not show up in these estimates, which have the sole purpose of comparing the trend followed by different types of wages.

Scattered data found on the remuneration of aparceros seem

to fit quite adequately to the figures shown by Table 13, but also point to the sharp fluctuations implied in this type of remuneration. In 1865, a year of relative price bonanza, at the estancia of Samuel Wheeler in Navarro, while peons were earning an average of 100 golden pesos per year, medianeros cashed between 280 and 390 golden pesos a year besides their increase in stock. Four years later, in the middle of the crisis, in Newton's Vista Alegre tercianeros were receiving between 115 and 220 golden pesos a year plus increase in stock; while shepherds on wages were paid 140 to 170 golden pesos per year.[30]

In view of the rising cost of sharecropping vis-à-vis wage labor, there was a gradual shift from *mediería* to *tercería* and *cuartería*, and an increasing proportion of the hands employed were hired as wage laborers. Furthermore, when the paddock system was introduced and the ratio of labor to sheep diminished about 50%, it was no longer profitable to engage sharecroppers in sheep raising, and most estancieros preferred shepherds on wages.[31]

The second question posed by aparcería, i.e., that of the danger of excessive independence of the sharecropper in relation to the central organization, was checked by compelling aparceros to hand in all the produce of their flock, thus removing them from commercial transactions. The central administration received all the goods, sold them, and paid aparceros their share. However, as we have seen in chapter 3, in the first decades of pastoral expansion, the worker enjoyed a certain independence in this respect, particularly in regard to the share of livestock he received at the end of the year. He could sell the livestock, or he could keep it to build up his own flock in order to enter other types of engagement which required capital in the form of sheep. By the 1880s, however, even this bit of autonomy was gone, as the sharecropper was no longer permitted to remove his share of the increase, and instead was paid for it at a fixed rate.[32]

The sharecropper became a wage laborer paid according to his productivity and that of his family, and to the prices of the staples in the market. Therefore, although his reward was generally higher than that of a hireling, he had to work harder with all his family to ensure maximum results and could never be sure of the income he was to receive at the end of the year, because it depended entirely on the yield of his flock and on the prices paid for his produce. As Gibson remarks, "If he earns a large sum of money one year, it is because his employer has made a proportionally larger one."[33]

In view of the advantages and disadvantages posed by this type of aparcería in relation to wage labor, most estancieros chose to combine both forms of engagement of the labor force, hiring, as we have seen, workers on halves, on thirds, and on fourths without capital as well as puesteros and peons on wages.

But these were not the only type of contracts entered into by sheep estancieros, as we have also found cases of the second type of aparcería, where the worker was a minor partner in the enterprise, as he contributed not only labor force and paid his part of the regular expenses, but also brought in capital in the form of sheep. We have seen that in this case the relationship between aparcero and estanciero resembled that of the tenant and the landlord, and while the former provided the labor force, the latter supplied the land (see above, chapter 3).

When the owner of the establishment could not afford to stock his land in such a way as to ensure the profitable exploitation of it all, he could choose to lease part of his estate, or he could introduce aparcería of this second type. Actually, estancieros sometimes even combined these two forms.[34] Leasing the land brought in a regular income which was far from insignificant through the whole period, but it frequently meant independence of the leaser from the owner in the way the productive process was conducted, although not always in the commercialization of his staples. Aparcería, on the other hand, ensured him a reliable shepherd for his part of the flock, and brought the puesto under the sphere of control of the central organization of the estancia. Even though these aparceros enjoyed more independence than sharecroppers of the first type in the way they managed flocks and disposed of their produce, they were nevertheless subject to the general rules of the estancia, had to participate in annual events such as shearing and curing of the scab, and quite often had to sell their shares of produce through the head station.[35]

In terms of income and costs, the prices of leases went up at a higher pace than those of labor force, and if in the 1850s and 1860s it was more convenient to give away the land to medianeros in exchange for their labor force, in the 1880s the share of aparceros was put down to one third, so that they increased their share of labor time dedicated to the estanciero's flock, while the share of land occupied by the aparcero's flock was in fact reduced.

With these different forms and characteristics, aparcería was a widespread practice among estancieros, who chose to employ shepherds under different types of engagement, including wage labor. As was mentioned above, they also hired permanent and temporary peons, as well as overseers and managers, all of whom were paid in wages. So were servants and cooks, but their role within the estancia was different from productive workers, as they did not contribute directly to the generation of surplus.

Briefly then, the estanciero sought the best way in which to achieve his main purpose: that of ensuring capital accumulation on the basis of his rural enterprise. He combined wage labor strictly speaking with aparcería in its different forms, sometimes even renting out part of his estate, thus realizing both profit and rent and playing the roles of capitalist and landlord, though generally subordinating this second aspect to the first. A complex system of social relations resulted within each unit of production, where the wage-labor–capitalist type of relation was predominant, but where other forms such as aparcería and domestic labor were found, and even the relation tenant-landlord could eventually appear.

The Economics of Sheep Estancias

The euphoria with which contemporaries viewed the prospects of the pastoral industry in the River Plate during its early decades of expansion and even until the deep crisis of the late 1860s was replaced by moderate optimism in the following two decades and by cautious concern at the end of the century. Although the industry prospered and expanded for over forty years in the province of Buenos Aires, for the individual capitalist the possibilities of making large profits with sheep declined throughout the period, and by the 1890s other fields offered better opportunities for investment.

In order to analyze this trend, I have constructed tables 14 to 17, estimating for different periods the costs involved in setting up a sheep estancia in the province of Buenos Aires, and the

Table 14
Initial Outlay of Capital Necessary to Establish a Sheep Estancia of
10,000 Ha in the Province of Buenos Aires (1845–1884)

Item	1845–1854 golden $	%	1855–1864 golden $	%	1865–1874 golden $	%	1875–1884 golden $	%
1. Land	8,700	19	64,000	57	100,000	58	160,000	62
2. Stock	34,320	74	40,580	37	60,800	35	64,960	25
40,000 sheep "al corte"	20,000		25,200		42,800		45,200	
800 rams	4,320		5,040		6,000		6,560	
400 horses and mares	4,000		4,000		4,000		4,000	
1,500 sheep, mestizo merino	4,500		4,755		6,000		6,900	
50 fine rams	1,500		1,585		2,000		2,300	
3. Physical assets	3,380	7	6,860	6	10,650	7	34,900	13
Central building (houses of owner and manager, kitchen, wells, etc.)	1,000		2,000		3,000		5,000	
Sheds	400		800		1,500		1,500	
20 puestos	1,000		2,000		3,000		5,000	
21 corrals	380		960		1,050		2,000	
Ditch or wire fencing	200		400		1,800		20,000	
Orchard	250		500		800		1,000	
Tools, carts, etc.	150		200		500		400	
Total	46,400	100	111,440	100	171,450	100	259,860	100

Table 15
Annual Expenses and Income of a Sheep Estancia of 10,000 Ha
in the Province of Buenos Aires (1845–1884)

	1845–1854		1855–1864		1865–1874		1875–1884	
	golden $	%	golden $	%	golden $	%	golden $	%
I. EXPENSES								
1. Wages of permanent workers	4,585	62	5,667	66	6,754	63	6,897	57
20 puesteros[a]	3,740		4,640		5,460		5,230	
3 peons	293		379		478		563	
2 foremen	228		252		336		460	
1 manager	240		300		360		500	
1 cook	84		96		120		144	
2. Temporary workers (except shearing)								
300 day-men	231	3	246	3	321	3	330	3
3. Shearing (including labor)	1,064	15	1,070	12	1,310	13	2,075	17
4. *Almacén*	400	5	450	5	450	4	500	4
5. Sundry expenses (including curing sheep, repairing tools, etc.)	1,000	14	1,000	12	1,500	14	1,700	14
6. Taxes (*contribución directa*, brands, *guías*)	50	1	200	2	350	3	600	5
Total	7,330	100	8,633	100	10,685	100	12,102	100
II. INCOME								
1. Sales of wool[b]	9,120	52	16,836	62	17,160	52	22,570	56
2. Sheepskins (approx. 5,400, or 450 dozen)[c]	900	5	1,080	4	1,800	5	2,700	7
3. Increase in stock (including sales as stock, for tallow, for butcher, and stock kept)								
Average 25%	7,565	43	9,129	34	14,180	43	15,215	37
10,000 sheep	5,000		6,300		10,700		11,300	
200 rams	1,080		1,260		1,500		1,640	
375 fine sheep	1,125		1,189		1,500		1,725	
12 fine rams	360		380		480		550	
Total	17,585	100	27,045	100	33,140	100	40,485	100

[a] Estimates based on a combination of wage labor and sharecropping, as follows: 1845–1854, 5 shepherds on halves, 10 on thirds, and 5 on fourths; 1855–1864, 10 shepherds on thirds, 5 on fourths, and 5 on wages; 1865–1874, 5 shepherds on thirds, 10 on fourths, and 5 on wages; 1875–1884, 10 shepherds on fourths and 10 on wages. Data on costs as per Table 13.

[b] Price of wool estimated in golden pesos per 10 kg, as follows: 1845–1854, 1.60$ (yield: 1.38 kg of wool per sheep); 1855–1864, 2.76$ (yield: 1.47 kg per sheep); 1865–1874, 2.60$ (yield: 1.60 kg per sheep); 1875–1884, 3.05$ (yield: 1.80 kg per sheep).

[c] Number of sheepskins results from adding estimated animals consumed in the establishment (which are not included in the 25% net increase of stock) and sold for tallow and for butcher. Prices estimated in golden pesos per dozen, 10-year averages: 1845–1854, 2.0$; 1855–1864, 2.4$; 1865–1874, 4.0$; 1875–1884, 6.0$.

Table 16
Profits Estimated for a Sheep Estancia of 10,000 Ha
in the Province of Buenos Aires (1845–1884)

	1845–54	1855–64	1865–74	1875–84
1. Total capital (in golden $)				
Initial outlay	46,400	111,400	171,450	259,860
Circulating capital	3,665	4,317	5,343	6,051
Total 1	50,065	115,757	176,793	265,913
2. Benefits (in golden $)				
Income	17,585	27,045	33,140	40,485
Expenses	7,330	8,633	10,685	12,102
Total 2	10,255	18,412	22,455	28,383
3. Annual profits as percentage of total capital (%)	20	16	13	11
4. Annual profits as percentage of total capital, excluding value of land (%)	25	36	29	27
5. Increase in average value of land between ten-year periods (%)	22	5	5	

Table 17
Profits Estimated for a Sheep Estancia of 10,000 Ha on Rented Land
in the Province of Buenos Aires (1845–1884)

	1845–1854	1855–1864	1865–1874	1875–1884
1. Total capital (in golden $)				
Initial outlay[a]	37,700	47,440	71,450	99,860
Circulating capital	4,013	6,876	9,342	12,451
Total 1	41,713	54,316	80,792	112,311
2. Annual benefits (in golden $)				
Income	17,585	27,045	33,140	40,485
Expenses[b]	8,026	13,753	18,685	24,902
Total 2	9,559	13,292	14,455	15,583
3. Annual profits as percentage				
of total capital (%)	23	24	18	14

[a]Excludes cost of land.
[b]Includes annual rent estimated at 8% of the average market price of land.

annual expenses, income, and profit expected for such an enterprise. These estimates are intended to be representative of the economics of a large number of those establishments that were in operation during the period and area under study, but I have found no way of assessing in a statistical manner the validity of this supposition. I have based this reconstruction on information collected from different sources,[36] and have tried to check their accuracy with concrete examples of estancias' modes of operation found in primary sources.[37]

A few remarks are necessary in relation to these tables. First of all, I have divided the period 1845–84 into four ten-year intervals which very roughly correspond to the successive stages of pastoral development in Buenos Aires: an initial phase in the 1840s and early 1850s; the great expansion of the late 1850s and early 1860s; the critical phase of contraction and readjustments which lasted till the mid-1870s and finally, the decade of maturity which inaugurated decline in the 1880s. These stages do not really fit into our ten-year intervals, but it was necessary to consider periods of equal time length in order to make averages comparable.

Secondly, in regard to size, I have considered only the large estancias, although I shall make side references to smaller establishments. Because most sheep were raised in estates of over 5,000 hectares, but few in those over 30,000 hectares (see above), I have chosen 10,000 hectares as the typical size for a large-scale unit of production.

In the third place, these estancias are entirely dedicated to sheep raising and wool production, as within their premises they do not engage in agriculture or cattle raising. Furthermore, they exclude _ activities such as breeding in cabañas or tallow making (*graserias*). They are run by the landowner, and the case of estancias on rented land will be treated separately.

As to labor employed, I have calculated costs assuming a combination of wage labor and sharecropping of the first type, according to the proportions more frequently observed by estancieros for each period. As the cost of sharecropping is included among the regular expenses of the establishment, it is not necessary to discount the corresponding shares from the estimates on income.

Throughout the period, I have assumed land of the same quality and carrying capacity (4 sheep per hectare); stock of similar quality (that which in each period was considered the average *mestiza* sheep), divided into flocks of 2,000 head; livestock with net reproduction rate of 25% per year,[38] and a labor force of equal qualifications.

Furthermore, I have considered that wool and skins sold at the same and uniform price for each period, not taking into account the different qualities that within one establishment always arose (*borrega* wool, belly, etc.), nor the better quality of the produce that might accrue from persistent mestization. The average yield considered was 3 lbs per fleece in 1845–54, 3.2 lbs in 1855–64, 3.5 lbs in 1865–74, and 4 lbs in the last decade.

In regard to physical assets, we have included for each period what we found to be the "average" investment.[39] Stock in these establishments has been estimated at 40,000 mestizo sheep, 800 rams, and a flock of fine crossbreed merinos of 1,500 sheep and 50 rams.

I have supposed that permanent personnel for such an estancia comprised a manager, two overseers, twenty puesteros (including wage laborers and sharecroppers), one shepherd in charge of the fine flock, one peon, one quintero and a cook. Temporary jobs outside shearing have been estimated at 300 man-days, while manpower involved in shearing is included in the total costs of that operation, calculated on the basis of cost per 1,000 sheep shorn.

Profits have been calculated as a percentage of total capital (fixed plus circulating capital, the latter estimated at half the annual expenses). These calculations do not take into account the perspective of the individual entrepreneur and therefore, in deducting expenses from income, I have not included rent of land nor interest in capital invested in sheep estancias. From such a point of view, only in the case of an estanciero owner of the land and in possession of all the necessary capital to establish and run his enterprise without resource to external financial sources, rent and interest should not be discounted from returns. Yet most estancieros at one time or another made use of credit and had to pay interest upon at least part of the capital in operation (see chapter 7), and those who leased the land upon which they established their enterprise had to pay as well an annual rent to the owner of that land.

Finally, let me add that data of prices and costs included in these tables were collected from different sources.[40] As no reliable long-time price series was found for any of the items involved in these calculations, the results are necessarily only an approximation to what the real figures might have been in the different cases. All prices represent averages over a ten-year period.

How do these figures compare with data from real cases and contemporary sources? Although I have not found complete accounts on any particular estancia for a reasonable number of years, in the course of my research I have encountered inventories of establishments of different sizes and for different years, as well as a few accounts for short periods of up to three years.[41]

As regards the capital outlay, the information available shows that within each particular period the proportion between the value of land, stock, and physical assets was quite similar for estancias of different sizes located in different partidos.[42] The same is true in the case of composition of expenses and income, although data in this case are rather scanty. These figures, in turn, are not too different from those included in our tables.

As regards estimates on profits, these are harder to check with primary sources as no long-term accounts are available to make the correct calculations. But I found a number of estimates of this kind in different contemporary sources, and after adjusting the figures according to a uniform criterion of accountancy the results do not differ too much from my own estimates, although these tend to be lower than the former.[43]

From tables 14 to 17 we observe the following:

(a) Throughout the period under study, the amount of capital required to put up an establishment grew steadily to represent in the 1880s over five times the amount necessary forty years earlier.

(b) Land and stock represented by far the most significant part of the capital invested. Both increased their value through the years, but land did so at a much faster rate than stock, so that if in the 1850s sheep represented 74% and land 19% of the value invested, by the 1880s these proportions were 25% for stock, and 62% for land. In the years following 1885 and up to the crisis of 1890 this tendency was further accentuated, making it less and less profitable to engage in an enterprise of the type herein considered. We shall see below how these problems were faced by estancieros and farmers.

(c) The price of stock rose steadily throughout the period (see Figure 5), but the pace of this increase was faster during the years of great expansion of the business (late 1850s and 1860s). This rise was the result both of a real increase in the value of sheep due to crossbreeding and the boom in demand as estancieros and farmers were stocking their lands with mestizo breeds. The situation was stable in the years that followed. Mestization continued, but at a slower pace, with no qualitative jumps until the 1880s, with the introduction of the Lincoln breeds and the *desmerinización* of the flocks as a response to increasing demand for mutton. Demand for mestizo sheep also slowed down, but it was kept at a level high enough to absorb the reproduction of the stock, thus avoiding the decline of prices in the long run.

However, periodic crises of the wool trade brought about a contraction in demand, generally accompanied by a drop in prices and an increased slaughtering of stock (see chapter 2). This helped to keep supply in check, and the prices show a stagnation from 1870 onwards. In fact, the curve of prices for common mestiza sheep runs almost parallel to that of wool prices. This should not

Figure 5

Price of Common Mestiza Sheep per Head
in Buenos Aires, 1847–1888

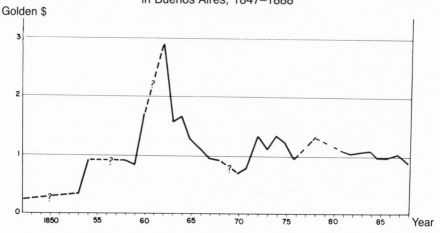

Source: Hilda Sabato, op. cit., App. V, table III.

surprise us. Until exports of mutton started to complement those of wool in the 1880s, sheep were raised mainly for the value of their wool; mutton, skins, and tallow were only byproducts whose relative importance depended entirely on the state of the wool trade and the price of the staple in the international market. Therefore, if prices of sheep in Buenos Aires were determined by demand of stock, this demand in turn was a direct result of the prospects of wool production.

(d) Physical assets represented only a small proportion of the expenditures involved in the establishment of a large estancia. However, their relative importance increased in the 1870s and 1880s, particularly with the introduction of more expensive materials for building, such as brick, and with the diffusion of wire fencing.

(e) The tables on annual expenses clearly show that wages represented by far the largest item of regular expenditure. The natural conditions of the pampean soil allowed for more extensive practices than those of other sheep-breeding areas, such as Australia, for example, where even after fencing had been introduced in the 1850s, the number of sheep under the care of one shepherd was around 1,000, while 2 acres of land were required per head.[44]

Nevertheless, in the province of Buenos Aires, where one man was in charge of 2,000 head in 500 hectares, labor accounted for 60% of annual expenses. This fact led estancieros to try to lower the cost of labor by attracting immigrants in order to improve supply, and by incorporating new technologies, such as the paddock system, which raised the ratio of sheep per shepherd. This interest in lowering costs did not go far enough to jeopardize the quality of labor, however. On the contrary, while profits remained high estancieros were willing to pay relatively good wages in order to ensure the proper qualifications required for a certain job, and we have seen how widespread the practice of sharecropping became throughout the period. Thus the incidence of the cost of labor on total expenditures reached its highest point in the decade of great expansion (1855–1864), while the lower values correspond to the last decades considered.

(f) Income depended as much on the sales of wool as on those of stock. Throughout the period, income produced by wool was enough to cover expenses and leave a surplus, but high profits were the result of considering not only cash receipts from the sale of produce but also reproduction of stock (whether kept in the estancia or sold) and the increase in the value of land.

In constructing the tables, I have considered a uniform price of wool, taken for each period as the average for ten years of the market price in Buenos Aires as per Table 20 and deducting from it the estimated cost of marketing for the first stage (see chapter 6). In the case of stock, the net reproduction rate has been estimated at 25% proportionally for each type of sheep, and prices represent averages. Therefore, these tables openly reflect in a very general way the trend followed by income in sheep estancias, and they say little about the individual results of any particular enterprise.

(g) These considerations have to be taken into account when analyzing profits. Although net proceeds tended to rise throughout the period, the trend of profits was clearly downward. In the first two decades of sheep expansion, returns seem surprisingly high. Capital invested in the business yielded 20% yearly in the decade 1845–1854 and 16% in the following decade. To these figures we have to add the increase in the value of land, estimated at 22% in that period. And although capital rendered high interest also in other endeavors, sheep raising became the most promising field for investment. Compared to cattle raising, the alternative productive activity which rendered high profits, the pastoral industry implied low requirements of capital, relatively quick returns, and could be started on a smaller scale. Other fields' of investment were also very rewarding, as interest on capital loaned yielded up to between 12 and 18%, and sometimes climbed to 24%. Commercial concerns were always quite profitable but very risky and capital-demanding.

The extensive scale of production in terms of land and labor, low prices of land, and rising prices for wool in the international market and stock in the local market all ensured the high returns shown for the first decade while the rapid increase in the value of land was responsible for the slight decline in profits—as percentage of total capital invested—experienced during the second period. The continuation of this tendency in land prices, together with the stagnation of prices for wool and stock were the main cause of the downward trend of profits in the following two decades. Actually, taken as percentage of total capital excluding land, profits rose during the decade 1855–1864, returning in the following periods to the level of 1845–1854. These results show the influence that the rise in the prices of land had on the economics of these estancias as it produced a sharp increase in total capital invested.

By the 1880s, returns resulting from sheep raising were far from being exceptional, and better profits could be found in other fields. In rural pursuits, these profits could be sought by introducing new productive activities and certain technical innovations. Thus, most estancias started to shift their main interest from wool production to stock raising for the purpose of selling it on the hoof to provide mutton for export. This shift brought about a rapid crossbreeding process—the Lincoln breed being preferred to the merino types.

Expansion of beef consumption in the internal market and the possibility of future exports to Europe led estancieros to invest in refining cattle within their establishments, thus incorporating a new branch of production in their enterprise. By the end of the century this process would be completed by the combined exploitation of cattle and agriculture, and the displacement of sheep to other areas. But in the 1880s we find that the most successful estancieros were crossbreeding sheep with Lincoln rams, and introducing fine cattle into their establishments. In their accounts, sales of stock became an important item.[45]

These long-term trends in profits do not reflect the fluctuations to which the pastoral industry was subjected throughout the whole period. Contemporaries became well aware of these sudden changes, particularly after the crisis of the late 1860s, and tried to introduce measures in order to control the effects of such fluctuations. Risky situations seem to have been more frequent and perilous during the first decades of expansion than in the following years. The most serious of all crises was that of the late 1860s, when after a few years of boom the downward fall was prolonged and steep. Profits, which had gone up to 30% or 40% in the days of bonanza, dropped to under 10%.[46] The situation seems to have become less dramatic in the following years, however. Price fluctuations continued during the 1870s and 1880s, but they were never so sharp as in the previous decades, and profits became more stable. In the Plate area, producers had devised a series of methods to check the effect of crises and natural catastrophes.

Already in the 1840s tallow making had been a response from producers to falling wool prices, but in the 1860s export of sheepskins and, later, sales of stock for human consumption had complemented wool and allowed for a certain amount of compensation in case of a decline in the demand for that product. In fact, prices for the staple in the international market were also more stable, if stagnant, because in the European textile industry the days of euphoria of the mid-century had long gone by.

Furthermore, estancieros had acquired the technical knowledge that enabled them better to control the effects of natural forces. For example, mortality of the flocks due to scab was lowered by regularly bathing the sheep in a special solution, and other diseases such as foot rot, throat worm, and fluke were kept in check by adopting preventive measures. Water conservation devices were improved to alleviate the problems caused by droughts, and more experience in flock management helped to obtain better yields and greater reproduction rates.

Besides the long-term trend of profits, and the short-term fluctuations within that trend, returns varied greatly between different individual enterprises, both in amount and in stability. For each estancia, profits depended upon annual income and expenditures, and the latter, in turn, varied according to the cost of labor and other inputs. These prices for labor costs and other inputs, however, were seldom subjected to the short-term sharp fluctuations that were typical of another item of expenditure: interest rates. We did not include them in our calculations, but they must be considered when referring to individual enterprises. Moreover, while market prices for labor and other inputs were probably very much the same for all estancieros, interest rates were weighed differently in each individual case, on the basis of the amount and type of credit each estanciero was able to obtain. For example, the ownership of more or less land meant differential access to credit, as mortgages on land were a traditional source for loans.

As regards income, it depended very much on the market prices of wool and other pastoral produce, and these prices suffered sharp variations throughout the period under study. For each individual entrepreneur, his income also depended on the quality and quantity of produce, and in this respect establishments that were well equipped and efficiently managed could show better results than more traditional enterprises. Probably, they were also better prepared to cope with market fluctuations and with problems caused by natural phenomena, thus achieving higher stability of income. Yet all these advantages were visible only in the long run, while sharp short-term differences among estancieros arose when selling the produce, as prices paid by merchants varied greatly in each case. Thus, a key factor in defining income was whom to sell to and under what conditions.

Thus, profits were far from being uniform for all establishments and their magnitude depended not only or primarily on the way production was organized but also on how commercialization was carried out and financing achieved. Thus, at the level of an individual establishment, while benefits derived from economies of scale or technological innovations in the process of production were slow to emerge and somewhat limited in nature, a strong connection with, or even direct participation in, the commercialization and financial networks could ensure the best possible returns for each transaction.

Under these circumstances, smaller establishments were more vulnerable than larger estancias. First, they were less inclined to introduce technical improvements: both in terms of equipment (wire fences, sheds, baths, etc.) and of stock (purebreeds, imported rams), the incidence of the cost of any one of these innovations on total expenditures was higher for these establishments than for the larger ones. Secondly, lesser estancieros were weaker vis-à-vis commercial and financial networks, and generally they received the lowest prices for their produce and paid the highest interest rates. All these factors affected negatively the amount and stability of income—a situation which was aggravated in times of crisis.

The periodic crisis suffered by the pastoral economy had deep repercussions at the level of the individual enterprises. Sudden drops in the price of produce, high mortality of flocks due to natural phenomena, or other such circumstances could lead to negative returns, and in the case of estancieros who depended entirely on their pastoral enterprise, this could prove fatal to their business. Crisis hit harder upon smaller establishments. For example, it has been said that the critical years of the 1860s brought about the liquidation of relatively small enterprises that presumably were swallowed up by larger estancieros or capitalists from other fields who could afford to invest and took advantage of the depressed prices of land to do so.[47] Although this process probably took place, it is hard to assess to what extent it affected the business. The deteriorated state of the trade probably led pastoralists to borrow money in order to meet their commitments, and when interest rates were high and the crisis prolonged, some of them found no way out of the situation but by selling out.

This relative disadvantage of some estancias vis-à-vis the establishments that were better equipped and capitalized and better related to the commercial and financial circuits would also prove decisive when structural changes swept the rural world: in the 1880s, when it became necessary to crossbreed on a large scale, and at the turn of the century when cattle breeding and agriculture replaced sheep in the area under study.[48]

(h) In Table 17, I have estimated profits accruing to tenant estancieros which obviously depended very much on the rent paid in each case. Already in 1853, MacCann stated, "Land is frequently rented for sheep, but no uniform rate can be named, as the value varied with the locality, and, in many instances, with the idea of the owner; some proprietors exacting double what others are willing to take."[49] These different rates coexisted with rates fixed by the government in public lands on lease, so that rentals were far from being uniform, even for land of the same quality and in the same year. These differences seem to have become less significant in the last decades of our period, and sources generally agree that rents were estimated at 6% to 10% of the market price of land.[50]

Calculations in Table 17 have been made on the basis of considering 8% as the average rate paid for rents throughout the period, in order to compare profits received by tenant estancieros and by estancieros who ran their enterprises on land of their own. For most of the period, profits accruing to the former were higher than those received by the latter, and even more so in the case of larger estancias, where the incidence of rents on expenses is lower than in the case of smaller ones. However, net profits were not the only factor taken into account by pastoralists, and renting land seems not to have appealed very much to these rural capitalists.

The practice of renting land upon which to establish a rural enterprise had a certain tradition in the River Plate. Most well-known estancieros in the second half of the century had started in the 1820s and 1830s as enfiteutas[51]—that is, renting public land—and the census of 1854 shows that leaseholders surpassed owners among estancieros.[52] This picture should not lead us too rapidly to wrong conclusions. First of all, renting of public lands has to be considered as a very particular kind of lease, as it was generally the only possible way of having access to that land, and of acquiring first rights to its acquisition when it was put up for sale. Secondly, if we go through the names of well-known estancieros and large sheep estancias, we find that by the 1880s most of them ran their establishments on lands of their own, and when they leased, it was generally on public lands.[53] Actually, I have found only one case of an important pastoralist renting private land in the area under study. I am referring to Juan Hughes, who, in society with his son, rented land in Rojas, Ranchos and Las Heras, on which he established large sheep estancias.[54]

In spite of these facts, in different sources mention is repeatedly made of rents and leases of lands devoted to pastoral pursuits. Unfortunately, few official censuses and statistics bring data on this practice, and when they do (provincial censuses of 1854 and 1881, and national census of 1895) they do not distinguish private from public land, nor land under tillage from that used to raise sheep and cattle. Thus, in 1854, the real incidence of rented land is probably distorted by land held in enfiteusis, and in the later sources distorted by the presence of agriculture. Furthermore, there is no reference to size or type of enterprise, and therefore there is no way of establishing the quantitative importance of land held on lease.

In spite of these problems, it is still possible to make a few remarks in relation to the practice of renting land in the pastoral industry. In the first place, it is quite clear from bibliographic and other sources that this practice was extended throughout the area and period under study. This does not mean, however, that it was desirable in the eyes of the pastoralists. "By all means, I would repeat, let emigrants purchase, instead of taking leases," insisted *The Brazil and River Plate Mail* in 1867.[55]

"Where lands are cheap and continually rising in value, no settler who can buy should rent"[56] was the recommendation of the same journal earlier that year. In 1881 a similar argument was stated in a report to the British Parliament: "The foreign sheep farmer frequently hires instead of buying land. Indeed, this system shows a better yearly profit, but larger gains, I think, are made in the end by buying, owing to the rapid increase in the value of good land."[57] Although the extraordinary increase in the value of land in the 1850s and 1860s was followed by a less impressive rise in the following decades, throughout the period land prices rose at an average annual rate of 8% in the area under study. Land was therefore a very secure long-term investment, which ensured capitalization; and its increased value, both real and expected, was considered by pastoralists as part of the benefits to be perceived by the business.

Other factors also accounted for the preference of pastoralists toward buying instead of renting land. It was again *The Brazil and River Plate Mail* that recommended: "Above all, let the emigrant avoid the common practice of taking a lease of land. He can only obtain a lease for three years; and the mere fact of his occupying it with cattle or sheep for that period will make it worth a great deal more; and, at the end of the three years, the occupier will either have to pay an enormously increased rental, or be turned out, and so have to begin again."[58] In effect, most leases where for from three to five years, after which—if renewed—they were updated in their prices, thus leading the tenant to a permanent situation of instability and uncertainty. Most likely, this consideration was taken into account more by small tenant estancieros— and farmers—than by large capitalist entrepreneurs, who were probably guided by the first argument in their decision to acquire land instead of renting it.

Finally, land became a source of credit. In effect, land could be mortgaged and this financial device, though used throughout the period to obtain capital, became most important after the creation of the Banco Hipotecario de la Provincia de Buenos Aires, in 1872.

On account of these advantages of possessing land in relation to renting it, most pastoralists were eager to become landowners, and when it was possible for them to acquire an estate, they did so even at the cost of becoming indebted and mortgaging their new property. When this step was not possible, however, pastoralists rented the land upon which to run their establishment. Generally, this situation was considered as an intermediate step toward full ownership of an estancia, even if it was not always achieved. Therefore, it should not surprise us to find that most large, well-known and prosperous estancieros were landowners, although smaller establishments—and especially farms— not infrequently were run by tenants.

The Process of Investment

To complement the rather static picture of the economics of sheep farming portrayed in the last pages, it is necessary to analyze the process of investment at the level of each individual enterprise and also at the level of pastoral industry as a whole. For the pastoral industry, its development implied (1) the utilization of newly available natural resources—mainly land and stock; (2) investment in land previously devoted to other activities; (3) the acquisition of pure and fine breeds imported from abroad to improve the quality of the native stock; and (4) investment in physical assets. Chapters 1 and 2 discussed the expansion of the industry over new lands and lands dedicated to other uses, so I will analyze here the process of investment in physical assets and fine stock. This process had recognizably different phases in the long way which led from a rather primitive, almost nomadic technique in the early decades of the century to the well-organized and equipped establishments of the 1880s.[59]

Commencing in the 1840s and going on into the 1850s, a first phase can be identified. In terms of physical investment (value in current prices of additions to durable physical assets other than livestock) this was the stage during which the spatial distribution of the old estancia was adapted to the new type of activity, establishing fixed locations for the puestos and its stockyards (fenced with boards called *lienzos*), drinking arrangements (water holes and wells), and minimal tools. The most important investment of this period was that which led to the introduction of imported rams and ewes to crossbreed with the native low-quality criolla sheep and thus obtain wools of better quality, which reached higher prices in the international market. In the late 1840s, of a flock calculated at 6 million head, MacCann estimated that at least one third were mestiza sheep.[60] By 1855, in thirteen counties north of the Salado River the total number of mestizas was higher than that of criolla sheep.[61] Interest in improving the local flocks had led as early as 1813 to the first introduction of merinos from Spain,[62] and again in 1824 and 1826 to new acquisitions from both Spain and England. In the 1830s a few pioneer estancieros were crossbreeding with fine sheep in their establishments and cabañas, with a total of 2,569[63] animals of different breeds being imported in 1837. The blockade of Buenos Aires apparently stopped this process, but it was resumed in the following decade with particular preference for merinos, especially the French Rambouillet, but also the German Negrettes.

Crossbreeding continued in the early 1860s, when the excellent prospects of the wool trade led to a rapid process of physical investment, forming a second stage in the process. Buildings in the estancia became more solid, particularly at the head station, where houses became more comfortable, and sheds, pens, and yards were improved to obtain better results. Shearing sheds, dipping yards, storage facilities—all contributed to reduce the average cost per unit of output and to improve the quality of the product. Planting of trees was also intensified during this stage.

The late 1860s brought crisis and a decline in investments. "Because of the depreciation of sheep and wool, all of us sheep raisers have no means of . . . introducing improvements," said the official organ of the estancieros in 1870. The article went on to say: "The only possible direction for us to follow is to fence. It is not a new idea, but if we can prove its cheapness . . . we think it is the only way of overcoming the present situation in the countryside."[64] And fence they did. Although wire fencing had been introduced in the 1850s, few estancieros had adopted it, with most establishments using only wooden boards to surround yards and folds, and a ditch to protect vital areas of their establishments. Fencing started in the early 1870s, but it was after 1875 that the process developed rapidly.[65]

Besides incorporating wire fencing, which reduced costs per unit of output, the process of physical investment was further intensified at this stage (late 1870s and early 1880s) by the expansion of brick housing, both in head stations and to a lesser extent in puestos. Together with improved techniques in the management of flocks, better built premises and more sophisticated tools and equipment led to an increase in the stability of output. Moreover, intensification of the crossbreeding process was heralding the new transformations of the pastoral industry which would take place after the mid-1880s.

The final stage brought about the gradual diversification of the productive activities within the estancia in response to the new demands: production of mutton and wool growing in combination became the main purpose of estancias, along with a growing interest in cattle. Investment was concentrated on acquisition of imported Lincoln breeds and on the rearrangement of the spatial distribution of stock within the estancias with the introduction of the paddock system. "The years 1888 to 1893," pointed out Gibson, "are witnessing the conversion of 50,000,000 sheep from one type to another [merino to Lincoln], a conversion probably without parallel in the annals of the sheep-breeding industry."[66]

The process of investment described for the pastoral industry as a whole ensured the expansion and transformation of production, thus contributing to the process of capital accumulation that took place in the area under study between the 1840s and the 1880s. However, this process cannot be explained by analyzing the sphere of production alone, and throughout this work I refer to other factors which must be taken into account in this respect.

For the individual pastoralist, investment was concentrated in three items: land, stock, and physical assets. Though the process was not an even one for all pastoralists and the relative importance of these items could and did vary in each case throughout the period, it is possible to trace certain trends in terms of investment at the level of the individual enterprise. When the pastoralist-to-be was not in possession of land (acquired previously for purposes other than sheep raising) or when the established estanciero wanted to expand his estate, he had to buy the necessary land either from the government or from private individuals. As we have seen, public lands in the area under study were released for sale in the early decades of the sheep expansion, and by the 1860s most land north of the Salado River was already in private hands. A relatively free market was then in operation, wherein most transactions in land were carried out. Direct transactions were widespread throughout the period, but the mediation of agents became more and more frequent in the 1870s and 1880s. Houses like Francisco Bullrich y Cia. became specialized in the auction and sale of real estate in general, rural establishments and stock.[67]

In order to stock his lands, the individual pastoralist had to acquire the common mestiza sheep to constitute his flocks, and he could also choose to buy fine rams and ewes to crossbreed. He made his acquisitions in the market. Direct transactions and ferias were predominant in the case of common sheep, although intermediaries were beginning to play an increasing role. Pure-breeds were imported by merchant houses at the request of local estancieros, but once crossbreeding establishments (cabañas) prospered in the province of Buenos Aires, they provided most of the fine stock for the area, although imports still continued and even increased. As in the case of land, specialized agencies increasingly mediated in transactions in stock.[68]

As for the process of equipping sheep establishments, this was achieved by drawing resources from within each estancia, as well as from the market. Thus, under-employed station resources were used when permanent peons or puesteros were called to help in building a hut or repairing a fence, when local mud was dried to make the adobe blocks which served to build huts, or when bricks were baked with raw materials from within the establishment. On the other side, recourse to the market was a constant feature of sheep estancias in their process of physical investment. Not only were many of the materials used in buildings and fencing bought in Buenos Aires or even abroad (boards, wire, furniture, tools, etc.), but quite often temporary peons were engaged to perform particular jobs, such as digging a ditch or a well, building a hut, or baking bricks. In this way, investment accounted for the direction followed by accumulation within the enterprise. Its reproduction was the result of a combined set of conditions, however, some of which are discussed in other chapters.

How to Become an Estanciero

The late colonial period saw a number of well-to-do merchants and royal officials in Buenos Aires invest part of their capital in the rural hinterland of the city, where they acquired rights to land and settled the first estancias. By the 1820s, after a decade of economic difficulties for the province, finally the local elites found a way of taking advantage of the opportunities opened up by the Revolution and the liberalization of international commerce. By attending to the demand for goods that could be produced in the River Plate, they began a process of productive expansion that proved most rewarding, both for those involved in the business and for the provincial treasury when the balance of trade showed increasing signs of improvement.

The expansion of production was achieved by capital provided not only by established estancieros but also mainly by urban interests. For the mercantile elite of Buenos Aires, investment in the rural areas was an answer to the difficulties found in urban endeavors, where foreign merchants were competing successfully with natives in all branches of business.[69] However, although attracted by cattle raising, these families of the elite did not abandon their urban interests. Instead, most of them diversified their activities, thus showing certain traits of behavior that were later to prove basic features of the Argentine ruling class.[70]

In this way was local capital directed toward the production of staples for export, taking advantage of the increasing international demand for goods that had already a certain tradition in the local productive world. Therefore, if this did not drastically change the course followed by production in the hinterlands of the city, it contributed decisively to intensify and accelerate its development.

The drive to the cattle business was given by, and conditioned to, the international market. When the latter started to show a new interest, and wool became an attractive possibility, our estancieros, always alert, gradually shifted from cattle to sheep raising. Such was the case of well-known families like the Canos, Guerrico, Pereyra, and Unzué.[71] Most of them had followed the traditional course which led from urban activities to the rural world, and in the latter from cattle to sheep. It is interesting to note that in most cases, while cattle estancieros had kept some sheep in their establishments as a complementary source of profit, large pastoral concerns frequently raised a few head of cattle for the same purpose. When wool growing looked profitable and the demand for hides and salted beef declined, sheep became not the complement but the alternative. The same thing was to happen, by the turn of the century, with cattle, when beef replaced wool and mutton as the staple export.

Tracking down the origin of those who became well-known sheep estancieros in Buenos Aires, we find that besides this traditional course from urban activities to cattle and then to sheep, two other courses were followed by pastoral families-to-be.

One group was composed of those who skipped the second stage, going directly from urban activities to sheep raising. Actually, this was the case of most of the families of foreign extraction, immigrants with capital who had invested successfully in commerce and other urban activities after Independence, and who followed later the route of their predecessor native and Spanish families, seeking to invest in productive activities which could yield high returns. It was they who became the pioneers of the pastoral industry. Latecomers as compared to the local families of the elite, well-informed about market conditions and having close ties abroad, these immigrants were quick to perceive the benefits that could be brought about by wool growing, which had for them the additional advantage of being a field still free from competition. They were joined by other immigrants: those who came with a little capital to invest in the rural field, and found sheep raising to be the best possibility (as cattle raising required more capital, was slower in yielding returns, and was almost monopolized by native families). The names of Hale, Harratt, Oldendorf, Halbach, Sheridan, and Stegmann are among those of the pioneer pastoralists.[72]

Finally, some of the well-known estancieros had made their fortunes within the industry itself, toiling their way up from the lower echelons to reach the highest ranks. Stories abound about Irish or Scottish immigrants who came without a penny in their pockets, were introduced to the pastoral business as simple workers, and in a few years became wealthy and powerful estancieros. Families such as the Duggans, Caseys, Hams, or Kavanaghs are always cited to certify the point, and although the path they followed proved much more successful than that taken by the bulk of the early immigrants, it was nevertheless possible during the early decades of sheep expansion.[73]

From a more favorable starting point, estancia managers sometimes became prosperous sheep breeders. Newton and Reid are perhaps the best-known examples of this possibility, as their establishments were among the most modern of their time.[74]

Most well-known estancieros, however, did not limit their business to sheep raising and wool growing. In effect, I have mentioned how most of those who came from urban activities did not abandon them, but simply extended their interests to include cattle raising or wool growing. The inverse path was followed by those who became estancieros by toiling their way up within the industry: most of them once they reached that condition started to invest in other concerns. Thus, we find that a large number of the best-known pastoralists participated as well in commercial concerns, urban speculation, financial investment, and public and private companies (stock in gas works, banks, railway, etc.).

Examples abound for both situations, but the cases of the Unzué, Hale and Casey families may help to illustrate the possibilities of diversification. The Unzué family[75] can be traced back to the eighteenth century, when Francisco Unzué held colonial posts in Buenos Aires, and was probably a merchant as well. One of his sons, Saturnino, in the first half of the nineteenth century became a well-known merchant and hacendado, and when he died in 1853 his inventory showed a total wealth of over 350,000 golden pesos distributed as follows: two estancias, over 150,000 golden pesos; urban property, including a *barraca*, 18,000 golden pesos; goods stocked in the *barraca*, over 32,000 golden pesos; cash and bank deposits, 38,000; stocks and bonds, 2,500 golden pesos; debtors, around 100,000 golden pesos. On his part, he owed around 75,000 golden pesos to several creditors.

The eldest son of the family inherited part of this business, amounting to 2.5 million pesos, or 125,000 golden pesos, but at his death in 1886 his wealth amounted to over 7 million golden pesos, he having multiplied the original capital by almost twenty times. Although most of his wealth was concentrated in rural concerns, besides estancias and two quintas, his possessions included urban property (18 houses and several lots), stock and bonds, and obligations from several debtors.

The example of Casey is well known.[76] Lawrence Casey was an Irish immigrant who went into sheep raising, becoming the owner of an establishment in Lobos in the 1840s. His son Edward rapidly diversified his business interests, investing in horse breeding, land speculating, brokerage, and a number of other concerns. He was at one time director of the Banco Provincia and participated in the construction of the Mercado Central de Frutos and of the Barrio Reus (Montevideo). At the age of fifty-eight he went bankrupt, and many of those who had trusted their savings to him followed the same fate.

Samuel Hale[77] was born in Boston in 1804, and in 1830 he made his first visit to Buenos Aires as supercargo under an American flag. By 1853 he had established the house S. B. Hale and Co., exporters and consignees. When he decided to invest in productive activities, he went into sheep raising by organizing one of the most modern estancias of the time (Tatay in Carmen de Areco). By 1879 his business interests covered a wide range of fields. His assets amounted to 4.5 million golden pesos, divided as follows: 1.6 million in estancias (cattle and sheep) and rural land in the province of Buenos Aires, Santa Fe, and in Paraguay; 150,000 in urban enterprises (houses and lots, a sawmill in Buenos Aires, a land *lavadero* in Barracas, two-thirds of a saladero in San Nicolás); shares, stocks, and bonds, over 2 million; debtors, 650,000. At that date his liabilities were high: he owed 2.3 million pesos to different banks and commercial firms. Much of the property was transferred to the creditors in agreements which were probably similar to those entered upon between Hale and Co., and the Bank of London related to estancia Tatay.

Although most estancieros diversified, we also find examples of many who concentrated all their capital in rural concerns. Such was the case with the Girados, of Chascomús, G. Mooney, Peter Sheridan, the Lawrie family,[78] and many others whose names are unknown today.

Though it is rather simple to systematize *ex post facto* the three main paths followed by those who became successful, wealthy, and well-known estancieros in the pastoral industry, it is not so easy to identify the course taken by those who became pastoralists, but never reached the highest ranks within the industry, and whose names have been lost or mean little to us today. Some of them worked their way up within the industry in those days of extraordinary profits, others probably invested savings made in other activities, especially commerce (*acopiadores, pulperos, barraqueros*).

But did these lesser estancieros abstain from investing in other fields or did they emulate the more successful ones by diversifying their interests? Although we know nothing about most of these pastoralists, I have found quite a few examples which indicate that besides investing in the rural world by acquiring land and stock, these families also put their capital in other pursuits, by buying urban property to rent out, and participating in commercial and money-lending operations at the local level.[79] The risky nature of the business (i.e., production and market risks) led many of these estancieros to diversify as well when possible, to be ready to respond to market conditions, and to be open to new possibilities.

Whether they could or did adapt to the deep transformation which took place during the last decade of the century is a question that simply cannot be answered at the present time.

Large-scale estancieros or small pastoralists; entrepreneurs who diversified their investments or who concentrated capital; fortunes of urban or rural origins—the monolithic image of the estanciero seems to fall apart. It is possible, however, to build a complex and heterogeneous picture of these men, and here we have tried to trace some of the more typical stories, which point both to the common traits and to the differences found within this class.

In an attempt to interpret the role played by these estancieros, scholars like Ortiz have taken pastoralists to represent the modern entrepreneurs of the rural world as opposed to cattle raisers.[80] This interpretation seems difficult to sustain, as estancieros involved in pastoral activities could hardly be defined as a homogeneous or distinct group in confrontation to those who favored the old style, opposing innovation. On the contrary, although some of the first to embrace sheep raising were immigrants seeking to invest in the rural areas, not a few of those among the main promoters of pastoral expansion were well-established cattle estancieros like Jurado, Senillosa, Unzué, or Cano. This is not to say that sheep raising was adopted by everyone, or that its advantages were seen from the very start by all rural entrepreneurs. The possibilities offered by wool growing were questioned by many in the early days of pastoral awakening, but already in the 1850s all doubts were rapidly being cast aside and sheep raising was attracting capital from almost all the other fields of investment.

Moreover, large and well-known pastoralists generally kept or developed an interest in cattle. Thus, their sheep estancias in the areas suitable for wool growing were only part of their rural business, which often comprised as well cattle raising concerns in the southern districts of the province which were carried out very much on the old style of the Rosas era, and quintas or chacras in the vicinity of the capital. On the other side, estancieros exclusively devoted to sheep raising could be either modern entrepreneurs or very traditional ones. In fact, most of the lesser estancieros and farmers ran their establishments in quite a primitive way, being unable or unwilling to introduce innovations and changes in the organization of their enterprises.

Leaving aside this debate regarding modern and traditional estancieros, Tulio Halperin Donghi has stressed the role played by a small group of men who aspired to lead and represent the interests of the agrarian classes. They were among the organizers of the Sociedad Rural, which was created in the 1860s by rich estancieros and entrepreneurs who were active in the commercial and financial circles of Buenos Aires as well as in local and national politics.[81]

In the first few years after the creation of the Sociedad Rural, this association was led by a heterogeneous avant-garde whose main purpose was not just to "represent and defend the aspirations of the landowning class," but rather "to reveal to this class what those aspirations should be."[82] Thus there were two sides to the role that the association and its leaders had to play. On the one side, they had to represent the corporate interests of the estancieros vis-à-vis the state, the political elites, and the mercantile classes. On the other side, they had to contribute to the development of the "class conscience" of their own people. Their message rested upon old rivalries between the agrarian interests and those of other sectors of society. It identified the hacendados as a particular group within the propertied classes and prompted them to play their "natural leading role" among the rural sectors. However, while this elite worked to arouse the political aspirations of estancieros and landowners in order to build the power base necessary to achieve hegemony in a projected hierarchical and harmonious society, these groups did not seem ready to play the role. Probably, they perceived too well that the changes experienced in the social and economic structure of the rural areas left no room for the consolidation of the social order devised by their leaders.

The development of capitalism came together with a growing complexity of the social structure. The estancia was to lose gradually the key role it had played both in the social and productive organization of the countryside. Thus, the possibility of consolidating a social order centered on the deferential relations between two complementary classes became more and more remote, as new sectors expanded in the lower and middle echelons of society. On account of these changes, most estancieros probably chose to ignore the message of their would-be leaders—an attitude which Halperin Donghi considers sensible since their "political power derived more from the central position they had achieved in the national economy than from any leadership they could come to exert upon the subordinate classes of the countryside."[83]

However, besides this awareness of the changes that were affecting the social order of the rural areas, it was the transformations of the landowning class itself that were at the basis of the indifference shown by most estancieros in relation to the proposals voiced by the core group of the Sociedad Rural. An increasing stratification within the ranks of the landowners was parallel to the consolidation of a group at the top whose interests became more and more intertwined with those of the rest of the propertied classes. Probably this sector found fewer and fewer reasons to kindle old rivalries, and rather than assert their role as hacendados, they chose to become members of a renewed national elite which did not reproduce traditional cleavages.

This last proposition cannot be proved within the limits set by this study of the sheep estancieros of Buenos Aires province. However, the findings included in this chapter do put into question the traditional image of the landowning class. In fact, they tend to dissolve this concept by stressing the process of stratification among estancieros, as well as the formation of an elite in which rural and urban interests were tightly knit together.

Notes

1. For the main characteristics of cattle estancias in the first half of the nineteenth century in the province of Buenos Aires, see: Jonathan Brown, *A Socioeconomic History of Argentina, 1776–1860* (Cambridge, 1979); Horacio Giberti, *Historia económica de la ganadería argentina* (Buenos Aires, 1961); Tulio Halperin Donghi, *Revolución y guerra. Formación de una élite dirigente en la Argentina criolla* (Buenos Aires, 1972) and "La expansión ganadera en la campaña de Buenos Aires," in T. Di Tella and Tulio Halperin Donghi (eds.), *Los fragmentos del poder* (Buenos Aires, 1968); Diana Hernando, *Casa y familia: Spatial Biographies in Nineteenth Century Buenos Aires* (Ph.D. diss., University of California, Los Angeles, 1973); John Lynch, *Argentine Dictator, Juan Manuel de Rosas, 1829–1852* (Oxford, 1981); Prudencio Mendoza, *Historia de la ganadería argentina* (Buenos Aires, 1928); María Sáenz Quesada, *Los estancieros* (Buenos Aires, 1980).

2. Herbert Gibson, *The History and Present State of the Sheep Breeding Industry in the Argentine Republic* (Buenos Aires, 1893), 29.

3. Among the pioneer pastoralists who engaged in sheep raising during the first half of the nineteenth century we may mention: Peter Sher-

idan (Irish), John Hannah (Scotch), Franz Halbach (German), John Gibson (English).

4. This classification is based on the categories defined by Archetti and Stolen in their book on the farmers of northern Santa Fe. However, unlike these authors, I shall not talk of farmer and capitalist modes of production, as I am referring to the internal organization of the units of production, whose main features cannot be generalized to classify social relations as a whole. (E. Archetti and K. A. Stolen, *Explotación familiar y acumulación de capital en el campo argentino* [Buenos Aires, 1975]).

5. Theoretically, under capitalist relations of production, rent is an extraordinary profit appropriated by the landowner upon whose estate a capitalist enterprise is organized by a capitalist entrepreneur who receives the regular profits. In this case, the estanciero is both the landowner and the capitalist, thus receiving both regular and extraordinary profits resulting from the business.

6. Already in the nineteenth century, politicians, such as Alberdi and Sarmiento, or technicians, such as Latzina and Martinez, theorized on the role of estancieros in Argentine society. More recently, the main interpretations on the role of estancieros have been put forward by historians and social scientists, such as Oddone, Ferrer, Puiggros, Pucciarelli, Halperin Donghi, Giberti, and Flichman. For a recent analysis on the subject as part of the problem of the Argentine ruling class, see Jorge Federico Sábato, *La clase dominante en la Argentina moderna: Formación y características* (Buenos Aires, 1988).

7. We have seen in chapter 2 that for a sample of sixteen partidos within the area under study, the structure of land property showed that in 1864 over 50% of the holdings were under 1,750 hectares, but most of the land (51.03%) was held in units of over 5,000 hectares, while in 1890, 63% of the holdings were between 100 and 1,000 hectares, but over 31% of the land remained in units of over 5,000 hectares. However, there were only two holdings over 30,000 hectares in 1864, and only three in 1890, setting therefore the upper limit for the kind of establishment we are interested in. These figures do not all correspond to land devoted to sheep raising, as units of under 1,000 hectares could easily have been dedicated to agriculture, especially in 1890. Furthermore, some of the larger estates could have been divided for rent, thus giving way to smaller units of production.

8. Wilfrid Latham, *The States of the River Plate* (London, 1868), 27.

9. For descriptions of pastoral establishments, see, among others, William MacCann, *Two Thousand Miles' Ride Through the Argentine Provinces* (2 vols., London, 1853), I, 70 and passim; Thomas Hutchinson, *Buenos Aires y otras provincias argentinas* (Buenos Aires, 1945), 316 and

passim; Latham, op. cit., 91 and passim; C. F. Woodgate, *Sheep and Cattle Farming in Buenos Aires, with Sketch of the Financial and Commercial Position of the Argentine Republic* (London, 1876); Emile Daireaux, *Vida y costumbres en el Plata* (2 vols., Buenos Aires, 1888), II, 187–190 and passim; Estanislao Zeballos, *Descripción amena de la República Argentina* (3 vols., Buenos Aires, 1881/88), III, 39–40 and passim; Carlos Moncaut, *Estancias bonaerenses* (City Bell, 1977) and *Pampas y estancias* (City Bell, 1978).

10. Sucesiones Francisco Girado, Elias Girado, and Federico Girado, AGN, Suc. No. 5993, 5995, 6119.

11. Daireaux, op. cit., 310–311 (my translation).

12. Gibson, op. cit., 69.

13. Ibid., 75.

14. Ibid., 137.

15. Ibid., 75–81.

16. See directories and annuaries of Buenos Aires and provincial towns, where estancieros are registered as residents (see Bibliography). In 1869, the National Census registered almost 12,000 estancieros in the province of Buenos Aires, excluding the capital city. Of these, only 7.6% were living in the scattered towns of the province, while the rest were all included as "rural inhabitants." In the city of Buenos Aires, the census registered only 565 estancieros, but surely estancia owners could also be found among those included as rentiers, merchants, and bankers. These figures may be misleading, as the census refers to *estancieros* and *hacendados* and these categories probably include not only estancia owners but also men involved in stock raising but not necessarily as capitalist entrepreneurs.

17. E. Daireaux, op. cit., 271 (my translation).

18. For a perceptive analysis of the articles published in the first issues of the *Anales*, see Tulio Halperin Donghi, *José Hernández y sus mundos* (Buenos Aires, 1985), chap. 6.

19. José Hernández, *Instrucción del estanciero* (Buenos Aires, 1964), 273 (my translation).

20. Sucesión Ricardo Newton, AGN, Suc. No. 7217.

21. *Archivo Senillosa*, AGN, Sala 7, 2-5-9.

22. On personnel employed within pastoral estancias and their duties, see, among others, Latham, op. cit., 26–28 and passim; Zeballos, op. cit., III, 303–320; Godofredo Daireaux, *La cría de ganado en la estancia moderna* (Buenos Aires, 1904), esp. 132–143 and 224–242; and Hernandez, op. cit., 238 ff.

23. On the combination of wage labor and sharecropping, see, among others, *Revue Britannique*, 1 (1851), 41; Latham, op. cit., 232; E. Daireaux,

op. ct., 311–312; Gibson, op. cit., 92–93; G. Cormani, *Argentina: Guida per l'emigrazione* (Milan, 1888), 131–133; G. Daireaux, op. cit., 135–136.

24. Sucesión Peter Sheridan, AGN, Sucesiones No. 8184.

25. Sucesión Henry Bell, AGN, Sucesiones No. 3971.

26. Sucesión Felipe Girado, AGN, Sucesiones No. 5976.

27. Sucesión Samuel Wheeler, AGN, Sucesiones No. 8760.

28. Sucesión Juan Acebal, AGN, Sucesiones No. 3695.

29. Suc. Federico Gándara, and Suc. Zemborain, AGN, Suc. Nos. 6119 and 8809.

30. Sucesiones Ricardo Newton and Samuel Wheeler, AGN, Suc. Nos. 7217 and 8760.

31. Gibson, op. cit., 274.

32. Ibid.

33. Ibid.

34. For example, in 1869, José Mariano Biaus' estancia in Chivilcoy (8,100 hectares) had fifteen tenants besides seven puestos in charge of shepherds on thirds and on wages. The establishment also employed several permanent and temporary peons, two cooks and two overseers. (Sucesión J. M. Biaus, AGN, Sucesiones No. 4026).

35. Gibson, op. cit., 92–93.

36. Estimates of costs, income, and profits of sheep estancias are found in MacCann, op. cit., I, 279–283; Augusto Brougnes, *Extinction du paupérisme agricole par la colonisation dans les provinces de la Plata, Amérique du Sud* (Bagneres-de-Bigorre, 1854), 120 ff.; Benjamín Vicuña Mackenna, *La Argentina en el año 1855* (Buenos Aires, 1936), 35 ff.; *The Brazil and River Plate Mail* (BRPM), 21 June 1864 and 22 April 1865; Hutchinson, op. cit., 316 ff.; Eduardo Olivera, "Nuestra industria rural en 1866," in *Miscelanea* (2 vols., Buenos Aires, 1910), I, 102–110; *Parliamentary Papers, Commercial Reports* (b) Embassy and Legation, vol. LXIX, 1867, 333; *Le Courier de la Plata*, 22 June 1866; Latham, op. cit., 221–226; M. G. and E. T. Mulhall, *Handbook of the River Plate Republics* (London and Buenos Aires, 1869), 15–18; Arthur Jerdein, *The Argentine Republic as a Field for the Agriculturalist, the Stock-farmer and the Capitalist* (London, 1870), 18–21; Richard Napp, *The Argentine Republic* (Buenos Aires, 1876), 307–309; Woodgate, op. cit., 24 ff.; E. Daireaux, op. cit., II, 326; Zeballos, op. cit., III, 39–42; Gibson, op. cit., 126; Archives des Affaires Etrangères (France) *Correspondance commerciale*, Buenos Aires (consular), vol. 11, 1879–1881, fs. 54–66 (3 June 1879); *ASRA*, vol. 3, 1869, 303–304, and vol. 16, 1882, 291.

37. Information on costs, income, and profits was found in primary sources, particularly in Sucesiones Unzué, S., AGN, Suc. No. 8578 of 1854; Sueldo y Gómez, B., Suc. No. 8184 of 1858; Sheridan, P., AGN, Suc. No. 8184 of 1861; Bell, E., AGN, Suc. No. 3971 of 1862; Girado, F.,

AGN, Suc. No. 5976 of 1863; Garahan, C., AGN, Suc. No. 5975 of 1863; Guevara, L., AGN, Suc. No. 5981 of 1865; Wheeler, S., AGN, Suc. No. 8760 of 1865; White, G., AGN, Suc. No. 8760 of 1866; Girado, E., AGN, Suc. No. 5995 of 1867; Girado, Fco., AGN, Suc. No. 5993 of 1867; Newton, R., AGN, Suc. No. 7217 of 1868; Biaus, J. M., AGN, Suc. No. 4026 of 1870; Acebal, J., AGN, Suc. No. 3695 of 1878; Lynch, P., AGN, Suc. No. 6624 of 1881; Gándara, F., AGN, Suc. No. 6119 of 1882; Girado, F., AGN, Suc. No. 6119 of 1882; Zemborain, A., AGN, Suc. No. 8809 of 1886; Sillitoe, T., AGN, Suc. No. 8333 of 1887; Stegmann, C., AGN, Suc. No. 8341 of 1888. Also in Libro de cuentas corrientes de Pastor Obligado desde 1-1-1849, AGN, Colección Biblioteca Nacional 800–801.

38. Sources mentioned above (note 36) estimate reproduction rates between 35% and 50% per year. Taking the lowest figure and accounting for losses and stock consumed, I have estimated a net reproduction rate of 25% per year, although rates probably increased with the improvement of stock-raising techniques.

39. In the case of the 10,000-hectares establishment, in the first period, the owner's house has three or four rooms, with walls of adobe and a tin roof; the manager has a similar house but smaller in size; and all puestos are primitive one-roomed huts. Shed and kitchen have no walls and their roof is thatched, while tools and furniture are scarce and primitive. The quinta is still in its beginnings, with a variable number of trees, and corrals throughout the estate are built in wood. Five thousand meters of ditch protect the key areas of the head station. In the 1860s the construction of a sheep estancia implied better housing than in the 1850s, wells of different types (*pozos de agua, jagüeles*) throughout the estate, more trees, and a better built shed for shearing. More improvements were required in the period 1865–1874, but significant changes were introduced mainly in the following decade. Wire fencing and brick housing were generalized, thus making it necessary to include them among provisions for cost of establishing an estancia.

40. See Hilda Sabato, *Wool Production and Agrarian Structure in the Province of Buenos Aires, North of the Salado, 1840s–1880s* (Ph.D. diss., University of London, 1981). Table 1, chap. 2; Table XII, App. IV; Tables III, IV, V, App. V.

41. For this information see Hilda Sabato, op. cit., Table II, App. V.

42. In some cases the percentage corresponding to physical assets was higher than the tables show, as certain estancieros invested more than others in equipment. But even in exceptionally well-equipped estancias like those of Newton in Chascomús (1872) and Tatay of Samuel Hale in Carmen de Areco (1879), physical assets did not account for more than 25% of the value of the establishment, while stock oscillated

between 30% and 38% and land between 50% and 56% (Sucesión New-ton, R., AGN, Suc. No. 7217, and *ASRA*, 13 [1879], 123, 155).

43. See Hilda Sabato, op. cit., Table I, App. V.

44. See G. Abbott, *The Pastoral Age: A Re-examination* (Melbourne and Sidney, 1971), 89–107.

45. For example in 1888, 75% of the income received by Stegmann from his estancia Poronguitos represented sales of cattle and sheep (Lincoln mestizas), and Gibson's calculations for the early 1890s include the acquisition of cattle and sheep (Sucesión C. Stegman [h], AGN, Suc. No. 8341; Gibson, op. cit., 126).

46. See, for example, Olivera, op. cit.

47. See "La Campagne," in *Le Courier de la Plata*, 28 November 1866.

48. Existing accounts of the lives of a number of estancieros bear witness to these changes. It is quite clear that entrepreneurs who proved most inclined to innovation during the pastoral age were later among the first to shift to cattle and grain production when mixed farming became more rewarding than sheep raising.

49. MacCann, op. cit., I, 286.

50. On price of leases and rent, see, among others, MacCann, op. cit., I, 8, 20, 86, 143; Hutchinson, op. cit., 310, 328; Latham, op. cit., 215; E. Daireaux, op. cit., II, 243–245; Zeballos, op. cit., III, 39–42; Gibson, op. cit., 60–67; *ASRA*, 19, 453 and passim, vol. 20, passim; BRPM, 7 September 1864; *Le Courier de la Plata*, 27 January 1867; *Censo general de la provincia de Buenos Aires, 1881* (Buenos Aires, 1883).

51. See chapter 2, and Jacinto Oddone, *La burguesía terrateniente argentina* (Buenos Aires, 1967).

52. *Registro estadístico del Estado de Buenos Aires*, 2nd semester 1854, Table 10.

53. For example, in the list of best-known pastoral establishments mentioned by Zeballos, almost all names correspond to estancieros who owned the land where they ran their enterprises (see Zeballos, op. cit., III, chaps. 5, 6, 7).

54. Sucesión Juan Hughes, AGN, Suc. No. 6361.

55. *BRPM*, 22 March 1867.

56. *BRPM*, 22 February 1867.

57. *Parliamentary Papers, Commercial Reports* (b) Embassy and Legation, vol. LXXXIX, 1881, 162–163.

58. *BRPM*, 22 March 1867.

59. See N. G. Butlin, *Investment in Australian Economic Development, 1861–1900* (Cambridge, 1964), 71–173, for an analysis of the process of pastoral investment in Australia.

60. MacCann, op. cit., I, 275.

61. The *Registro estadístico del Estado de Buenos Aires* of 1855 registers for those counties a total of 1,033,642 mestiza sheep and 856,361 criollas (1st semester, 46).

62. Mendoza, op. cit.

63. MacCann, op. cit., I, 278.

64. "La cría de ovejas en el Plata," in *ASRA*, 4 (1870), 131 (my translation).

65. Imports of wire for fencing show a rapid increase after 1870. See Zeballos, op. cit., 254–255.

66. Gibson, op. cit., 40.

67. See advertisements in current newspapers and journals, such as *La Prensa, La Nación, Anales de la Sociedad Rural Argentina (ASRA)*.

68. See names of importers, for example, in Zeballos, op. cit., 28, and Manuel Macchi, *El ovino en la Argentina* (Buenos Aires, 1974), passim.

69. Tulio Halperin Donghi, *Revolución y guerra*, 114–118.

70. See Jorge F. Sábato, op. cit.

71. See Hernando, op. cit., for the history of the families Cano, Guerrico, Pereyra, and Unzue.

72. On the history of these pioneers, see, among others, Mulhall, op. cit., edition of 1869, 45, 125–126, 146–150; Hernando, op. cit., 199–239; Mendoza, op. cit., 124–127; Gibson, op. cit., 13 ff.; Zeballos, op. cit., III.

73. J. C. Korol and H. Sabato, *Como fue la immigración irlandesa a la Argentina* (Buenos Aires, 1981).

74. On the history of Reid and Newton, see particularly Mulhall, op. cit., edn. of 1869, 135–136; Mendoza, op. cit., 162–164; *ASRA*, 3 (1869), 147 ff.; Gibson, op. cit., 196–200; Zeballos, op. cit., III.

75. Sucesiones Saturnino Unzué (AGN, Suc. No. 8578) and Saturnino E. Unzué (AGN, Suc. No. 8590). Also Hernando, op. cit., 510 ff.; Carlos Lix-Klett, *Estudio sobre producción, comercio y finanzas de la República* (Buenos Aires, 1900), 1202–1203.

76. Virginia Carreño, *Estancias y estancieros* (Buenos Aires, 1968), 270–280; Charles Jones, *British Financial Institutions in Argentina, 1860–1914* (Ph.D. diss., Cambridge University, 1973), 48–50; Eduardo Coghlan, *Los irlandeses* (Buenos Aires, 1970); Saenz Quesada, op. cit., 250–254.

77. Mulhall, op. cit., edn. of 1869, 45; AGN, Cédulas censales del Primer Censo Nacional de Población, 1869, Booklet No. 58, Carmen de Areco; *BOLSA*, Letter Book D 35, Buenos Aires Branch to Head Office, Confidential, letters of 11 July, 24 September, and 10 October 1879, and 24 April, 23 October, and 8 November 1880.

78. Sucesiones G. Mooney (AGN, Suc. No. 6852) and P. Sheridan (AGN, Suc. No. 8184). Also Hernando, op. cit., 199 ff.

79. Sucesiones G. Bell (AGN, Suc. No. 3971); L. Guevara (AGN, Suc. No. 5981) and S. Wheeler (AGN, Suc. No. 8760).

80. Ricardo Ortiz, *Historia económica de la Argentina* (2 vols., Buenos Aires, 1964).

81. Tulio Halperin Donghi, *José Hernández*, chap. 6.

82. Ibid., 225 (my translation).

83. Ibid., 262.

5

The Sheep Farm

There is as yet no middle class; the owners of land feeding
immense flocks and herds form one class, their herdmen
and shepherds form another: but immigrants are beginning
to form an intermediate class of small flock holders,
answering to our English yeoman.
—W. MacCann, *Two Thousand Miles' Ride Through the
Argentine Provinces*, 1853.

As MacCann expected, sheep farms became widespread in the
years that followed his statement quoted in the epigraph. A large
number of small enterprises devoted to sheep raising and wool
growing sprang up throughout the province of Buenos Aires,
particularly north of the Salado River, during the 1860s and well
into the 1870s. Their extension was frequently around several
hundred hectares, and they often belonged to immigrants who
worked hard along with their families to build up and enlarge
their enterprises.

Very little has been said about these farmers. Contemporaries
refer to them, but only briefly,[1] concentrating attention on estancias and their owners. Historians have carried this bias even further, and very few of their writings mention sheep farmers without
much more than just naming them.[2] In primary sources and documents they are usually hard to identify, as they belong to the
anonymous characters of history. To complicate matters further,
contemporaries did not give them a particular name in Spanish,
although those who wrote in English generally distinguished between estancieros and sheep farmers.

More than the presence of farmers in Buenos Aires, it is their
absence that is permanently mentioned by historians and social
scientists. In opposition to the pattern of development followed
by other new cattle- and grain-producing areas of the world, and

even of Argentina (for example, Santa Fe), where production was in the hands of farmers, Buenos Aires is seen as defining a different model, where large estancias set the pace of rural development.[3] Therefore—and quite reasonably—attention is concentrated on the predominant type of enterprise, and forms other than estancias are left aside in most studies. Even those authors who claim that latifundia were not that important after all,[4] fail to analyze the type of enterprise they seem to think expanded in the area at the expense of large-scale productive units.

Although it is possible to contrast a pattern of social development based mainly on the expansion of farms against that which results from the consolidation of large enterprises, this does not necessarily imply that farms and estancias were incompatible and that the expansion of any one of them excluded that of the other.[5] Actually, during the sheep era in the period and area under study we see not only the multiplication of estancias and farms, but also the polarization of land property together with its relative fragmentation. And while estancieros not only remained among the predominant economic forces during the period, but also exerted social and political power, farmers adopted a complementary role in production, and a secondary place in society. How and why they accepted this role and what was their political significance are still unanswered questions. In the following pages I shall turn to a more restricted aspect of this subject—that of sheep farming, its economic significance, and some of the problems it faced in its expansion.

The Expansion of Farms

We have defined sheep farms as enterprises whose main purpose was to produce pastoral goods to sell in the market at a profit, and which relied mainly on family labor, only occasionally or secondarily making use of wage labor. We also said that farmers were not peasants, because they were not only concerned with the survival of the unit of production and the subsistence of themselves and their families, but also and primarily with the reproduction and expansion of their enterprises. Farmers owned the means of production within their enterprises, including the land, although they frequently contracted debts to acquire such means.

Tenant farming was also a widespread practice, but as explained earlier I shall deal with it separately. The farmer bought his means of production as well as goods for the consumption of the household in the market, where he also sold his produce. Therefore, although his way of facing production was not strictly capitalist as it was not based on wage labor, he participated fully in the capitalist relations of distribution, and his enterprise could not be thought of independently of such a system.

As we see, a theoretical characterization of farms inevitably leads us to the long-lived discussion on peasantry, petty commodity production, and the relationship between these forms and capitalism.[6] Without denying the importance of that discussion, I wish to avoid generalizations on the subject and will thus concentrate on the analysis of the main features of sheep farms in the province of Buenos Aires in the light of that discussion while hoping that it will offer new evidence to support some of the arguments advanced so far.

We have already called the years between 1850 and 1880 a period of transition during which the basic features of Argentina as a country and a nation were defined. At the beginning of the period, however, although some of these characteristics were already present, others were only barely sketched and yet others were not even predictable. Although already set in the way of capitalism, it would be during this period that accumulation was achieved on a significant scale, and that social relations reached their full development. The class structure of the 1840s had still a long way to go to attain the complexity of forty years later.

The expansion of sheep raising and wool production represented an important step in this process of accumulation. In an area in which the relatively limited economic development of the previous decades had been monopolized by a class never too tightly closed to new members, the opening up of opportunities of such scope was accompanied by the incorporation of a goodly number of newcomers into the ranks of those who had control of the economic resources of the country. This process gave way not only to the enlargement, renovation and consolidation of the bourgeois class but also the expansion of a class of artisans, shopkeepers, small merchants, and landowners, which already existed in the previous decades, but found an unprecedented impulse in the period under consideration.

It is in this context of changes and readjustments in the social structure and of economic growth that the emergence and expansion of farms becomes possible. Within this context, certain specific factors and conditions contribute to explain how and why this type of enterprise did prosper in Buenos Aires of the pastoral era. These two problems are closely related to a third one: that of who became the farmers.

The first factor, which is an obvious prerequisite of this type of enterprise, refers to the technical characteristics of sheep raising and wool growing. In effect, due to the way in which pastoral production was carried out, one shepherd with his family was considered enough to take care of a flock of 2,000 sheep, which occupied around 500 hectares (or less) of land, and yielded enough produce in normal years to pay all expenses and leave some profit. Furthermore, quick returns could be expected in sheep raising, which—after at most one year—gave its produce in wool and stock. Therefore, given these technical conditions, small enterprises could be organized at profit with little initial capital.

A necessary condition for these enterprises to expand, however, was the availability of resources, mainly land and stock, which were the basic means of production for such establishments. As we have seen, although the land market was in operation during the whole period, it was during the first two decades that land was relatively cheap and more easily accessible to someone with little capital. Furthermore, until the 1860s investment in land represented less than 50% of the total needed to establish a farm (see Table 18). Stock, on the other hand, was also available in the market and although it was quite cheap in the 1840s and early 1850s, demand put its price up in the following decade, to remain stationary in the 1870s and 1880s. But stock was also obtainable by sharecropping, and many a farmer had acquired his sheep in the years that preceded his establishment as a farmer, when he had worked on shares for an estanciero. This way of having access to sheep was very much in practice in the 1850s and up to the 1860s, but later, as we have seen, it became more and more restricted. However, since the 1870s, stock represented less than

Table 18
Initial Outlay of Capital Necessary to Establish a Sheep Farm of 1,000 Ha
in the Province of Buenos Aires (1845–1884) (in golden pesos)

	1845–1854		1855–1864		1865–1874		1875–1884	
	Amount	%	Amount	%	Amount	%	Amount	%
Land	870	24	6,400	64	10,000	64	16,000	72
Stock	2,532	70	3,124	31	4,980	32	5,280	24
4,000 sheep	2,000		2,520		4,280		4,520	
80 rams	432		504		600		660	
10 horses	100		100		100		100	
Physical assets	216	6	442	5	630	4	940	4
Main house (includes kitchen,								
well, furniture, tools)	100		200		300		500	
1 hut	50		100		150		250	
2 folds	36		92		100		190	
Trees, orchard	30		50		80		100	
Total	3,618	100	9,966	100	15,610	100	22,220	100

30% of the investments made in the farm, while land's share had gone up to over 65%.

Thus, in the early days of sheep expansion, resources necessary to establish a farm in the area north of the Salado were relatively cheap and easily available. In the 1870s and 1880s their price rose in such a way that the farmer-to-be had to dispose of almost six times more capital in the mid 1880s than forty years earlier. Access to this sum of capital therefore also became a problem, and the issue of credit acquired an unprecedented relevance.

A third condition for the relative success of farming was the existence of commercial and financial networks which were open to the requirements of farmers. As we shall see, however, the relationship established between the farmers and those networks became at the same time one of the most serious sources of limitation for the process of accumulation within the sheep farms.

Finally, farms probably would not have come into existence were it not for those families who became the main source of labor power in that type of enterprise. Although labor requirements in sheep raising were relatively low, labor represented the main item in the regular expenditure of capitalist estancias. The replacement of hired personnel by household members actually made the farm a feasible enterprise. But family labor is not a universal practice, and its incorporation in the Argentine pastoral industry owes much to the immigrant settlers. Irish, Basque, and Scotch settlers were trained in peasant self-exploitation, and brought with them the values of societies in which family farming was an extended practice. Some of them, therefore, took full advantage of the possibilities offered by sheep farming and that is why most contemporaries equated farmers with immigrants—particularly Irish immigrants.

Not only Irish settlers became farmers, however, although they were probably among the first pastoralists to engage in this type of enterprise. The most frequent way to sheep farming was followed by rural settlers who, through wage-labor, sharecropping, or tenant farming, sometimes had the opportunity to buy land and acquire stock, thus initiating their own enterprise. Labor power was to be found within the family, which had already been well trained in pastoral jobs. As we have seen, this goal probably was achieved more easily in the early decades of our period, when land was still cheap, labor well remunerated, and returns high enough to ensure rapid capitalization.[7]

For nonrural settlers, becoming a farmer must have been more trying than for those who were used to life in "the camp." Nevertheless, probably more than once savings made in urban activities found their way to the countryside, and small artisans or shopkeepers became farmers.[8]

What was the fate of these farmers? Individually these families may have remained as farmers for a long time, even generations, but they could also have developed into capitalist estancieros, or even have left the countryside altogether in order to start a new life elsewhere. However, sheep farming considered as a manner of facing and organizing the process of production persisted throughout the period, as several conditions ensured the reproduction of the system.

Although the economic significance of this system as compared to that of estancias is hard to evaluate, I have estimated the proportion of land that could have been in the hands of farmers.[9] According to my estimates, and if we consider that in the period under study rural production was in direct proportion to the amount of land held, it can be concluded that sheep raising was predominantly carried out in estancias, as described in chapter 4, but that family farms shared this activity, covering a variable proportion of production which could not surpass 15% in the mid-1860s, but which probably reached a higher proportion in the following years, as the significant increase in the amount of land held in plots between 500 and 1,750 hectares, registered in most counties in 1890, could hardly be attributed only to the expansion of agriculture.

Operating a Sheep Farm

Sheep and land thus became the basis upon which a family could start a pastoral enterprise. The way in which the process of production was carried out varied slightly from farm to farm, but all of them shared certain basic features which enable us to classify them in a different category to that of the estancia. Although the main characteristic of farms was that they employed mostly domestic labor, other traits were associated with this type of enterprise as well.[10]

Most sheep farms were entirely dedicated to pastoral activities. Wool growing was the main item in the business, but stock raising was the necessary complement, as it became an important source of income when sheep were sold, and of accumulation when they were kept in the farm. Sheepskins and other by-products were only of secondary importance, except in years of high mortality caused by epidemic or provoked by the farmers to control overproduction. Unlike most estancias, however, farms very seldom kept cattle or horses for commercial purposes.

The homes of these farmers, and indeed their whole way of life, did not differ much from those of shepherds and sharecroppers employed by pastoral establishments. Within the boundaries of the farm, a house was generally built for the family, with kitchen, well, and corrals. A hut or two could be found in the periphery to board those members of the family in charge of a particular flock. Mud and thatch were the basic building materials, although in the 1870s brick was sometimes used in the main residence. Tools and other equipment were generally primitive and scarce, and fences were seldom found.

Everyday life was governed by the routine of work. But, unlike hired shepherds, these farmers worked for themselves, as the main source of labor on the farms was the family. The head of the household was at the same time owner, manager, and worker within his establishment, and each member of the family was assigned certain jobs to carry out. Besides the main couple, farm households could include several sons and daughters, single or widowed brothers or sisters, orphan nephews or nieces, one or more grandparents, and at times even an *agregado*, a stranger or a distant relative who was adopted into the family but had a subordinate place within the household. Thus, every farm always had a number of hands available for work. All the men, excluding only the very young and the very old, were in charge of the farm work, while women generally were responsible for housework and subsistence production although they could also help men in their pastoral activities.

Hired labor was employed only sporadically, particularly during the shearing season. Shearing was a collective task engaging not only the landowner, his family, and hired laborers, but other farmers as well, who helped each other during the season. But farmers employed wage labor as well on other occasions, generally for specific jobs, but sometimes hiring a peon for a longer period to help in the more regular tasks of the farm.

> With the exception of herding and shearing, all work in connection with the flocks is reciprocally gratuitous. When, for instance, it is requisite to dip the animals for "scab," or other disease, to mark lambs or new sheep, the owner usually makes the round of his neighbors, and seeks their assistance, which, as the service is always returned, is invariably granted. A day is fixed, and the host is bound to provide his helpful guests with wine, food and *mate*, which are ungrudgingly supplied. It is also customary to kill a fat sheep or lamb.
>
> On the arrival of the time for shearing, the sheep farmer engaged a gang of shearers . . . whose services are easily secured.[11]

Although the cost of establishing a farm of 1,000 hectares has been estimated in Table 18, calculations of expenses, income, and profits have not been made in this case as they would be too farfetched. The farmer did not consider self-remuneration as an expense, and therefore it would be hard to estimate the cost of labor in his enterprise. Domestic and farm expenses were together in his calculations, and profit and self-remuneration became inseparable. Income is also hard to appraise. Average yields would probably differ from those observed for estancias, but we have no way of estimating them. Also, prices received for produce would prove different. As we shall see, farmers commercialized most of their goods through middlemen, obtaining generally lower prices for them than estancieros.

But the problem of expenses becomes relevant when analyzing the question of capital accumulation and the reproduction of the farm system. Therefore, as quantitative data are not available, the subject will be approached through information found in different qualitative sources. Running a sheep farm required relatively little circulating capital, as current expenses were mainly composed of domestic expenditures[12] and by those required to operate the farm, which included only one significant item: shearing costs.[13] As sheep farming was a seasonal activity which yielded its main returns once a year, these costs had to be financed by resorting to credit, which was also required by the farmer to acquire the means of production to establish or enlarge his enterprise, or to ensure its survival in critical years. These loans bore a regular interest which has to be included among current expenses of the farm, as it probably represented an important cost for the farmer.

Farm income resulted mainly from the sale of wool and stock, and therefore depended entirely on the prices the farmer could obtain for his produce. In consequence, farm returns, considered as the annual income received by farmers after subtracting current expenses, exclusive of self-remuneration but including domestic expenditures and interest on loans, depended mainly on the extent of domestic consumption, the degree and conditions of indebtedness, and the price of produce. All three of these items, as well as their combination, could vary greatly from farm to farm. Thus any attempt to generalize on the performance of farmers is at least vague, if not void of any real significance.

Nevertheless, in the opinion of contemporaries at least until the 1880s, in any average year sheep farms could expect positive returns, not only ensuring the subsistence of the owner's family and of the enterprise but also allowing for extra income. However, the problem was that an "average" year seldom occurred, as the pastoral industry was subject to yearly fluctuations which affected small enterprises even more than estancias. So that one or two seasons of booming prices and extraordinary returns could easily be followed by a depressed period in which the family had to lower consumption to minimum levels in order to be able to meet expenses. Thus we find families of farmers who made fortunes, while others ended up bankrupt, with many more probably following a long course of very gradual and uneven development, slowly improving their enterprise, their living conditions, and even their social status.

In effect, positive returns opened the way to investments, and these in turn to capital accumulation. Although the farmer could and did invest in his enterprise, he did not neglect his household. Thus, while his first home was probably no better than the hut of a hired shepherd, he would gradually enlarge and improve the building, always within the limits imposed not only by his income but also by his values. In the case of Irish farmers, we have seen that returns sometimes were also channeled into education, making an effort to ensure schooling for their children, or to community action, donating money to different funds.

The farmer was always ready to invest in expanding his enterprise, and this he did by buying more land and stock, by refining his flocks, and by improving his tools and equipment. However, I shall argue that there were limits to his possibilities of accumulation. If farms had found no limitations to this process, they should have become only a stage in the way to pure capitalist enterprises. Yet the sheep farm was not a prelude to the estancia. On the contrary, different conditions contributed to limit that possibility, and to ensure the reproduction of farms as a distinct type of pastoral enterprise during the period and area under study. Although most of these limitations arose outside the sphere of production, the scale of the productive process and the type of labor force employed contributed to slow down the pace of capital accumulation and ensure the reproduction of the farm system.

From the point of view of the organization of the productive process, the mere size of farms limited the degree of technical improvement that it was profitable to introduce in the enterprise. As we have seen in the case of estancias, although physical assets represented only a small part of the total investment in these pastoral establishments, nevertheless economies of scale did take place, particularly with innovations in management and equipment introduced in the 1870s and 1880s. For a farmer, to wire-fence, to build a dipping plant, or to acquire a purebred ram was hardly justifiable, not only because the investment was relatively too high and could become difficult to amortize, but also because it would remain underemployed.[14] Furthermore, adoption of labor-saving devices would only make sense for the farmer if the existing relationship between size of farm and size and composition of the family did not allow for the optimum exploitation of resources.

Nevertheless, farmers did invest in improving stock, equipment and buildings, and even in fencing, and we find already in the 1870s farms which have crossbred flocks, wire introduced in fences to protect certain key areas of the establishment, sheds to carry on the shearing, and other such improvements.[15]

These improvements were directed towards obtaining better qualities and yields both of wool and stock and thus higher and more stable returns. But any significant increase in production implied also the expansion of the flocks, mainly by retaining the extended flocks which resulted from year-to-year reproduction, but also by acquiring new animals. This expansion, however, soon found a limit in the extension of the farm. Given the technical conditions of the time, the carrying capacity of the soil varied from place to place, but for any given establishment it had a limit. If the numbers of sheep per hectare were carried too far, overstocking ensued, and returns started to decline. Proper expansion, therefore, required the occupation of new land.

This matter proved quite difficult to solve. In effect, farms not only required land, but this land had to be contiguous to the primitive establishment. The organization of the productive process—in which all the family took part—could hardly be maintained if the farm were split into two parts, although there were cases in which a member of the family was put in charge of the new lot, which in fact generally meant a new unit of production.[16] The extraordinary increase in the value of land that took place in the period and area under study was accompanied by a parallel process of falling profits, therefore making it harder for farmers to buy more land in the same area.

Mere size made farms also very vulnerable to production risks, introducing a disturbing element into the process of accumulation as it diminished the possibilities of making a sustained effort in that direction. Epidemics, droughts or plagues could literally decimate the few flocks of the farmer, leaving him with little possibility of recuperation. Also, market risks had deep repercussions in farms, and practically all crises suffered by the pastoral sector seem to have hit farmers harder than estancieros. The crisis of the late 1860s in particular apparently caused the ruin of many of those pastoralists, who had been able to acquire land and stock in the years of boom, but could not overcome the problems brought by the crisis.

Further hindrance to the process of accumulation arose from the type of relationship established by farmers with commercialization and financial networks related to the pastoral industry. Farmers generally commercialized their staples through middlemen, obtaining lower prices for them than other larger producers, and becoming most vulnerable to market risks. Furthermore, in order to start his enterprise, to enlarge it, or to face current expenses, the farmer quite often had to resort to loans, and although he had access to different sources of credit, not infrequently he became too dependent on those who provided him with capital.

More limitations to the process of capital accumulation within farms resulted from factors other than those strictly economic. Questions such as the problem of the subdivision of properties imposed by the laws of inheritance as well as by custom, or the reluctance to hire wage labor as a consequence of certain sets of values the farmer had learned to cherish, should be considered matter for future research.

In spite of all these limitations, and with the ups and downs imposed by the fluctuating market for pastoral production, capital accumulation took place in the farming sector. Farmers sought ways in which to overcome those problems and by and large not only did they invest in their own enterprises but also looked for other fields into which to put their capital.

Thus, we find farmers incorporating an associate into the business so as to allow for the expansion of the enterprise without risking too much capital or having to resort to hired labor; renting land to enlarge the estate when acquiring it was not possible;

summoning some relative who might live in town, in another rural area, or even abroad, to come and work on the farm so as to enlarge the household labor force, and finding other innumerable ways of getting ahead with their farms.

Unable or unwilling to continue putting capital into their enterprises, some farmers sought other fields for investment, and we find them buying land to rent out, or urban property, particularly in provincial towns. Finally, some families of farmers were so successful in their process of accumulation that they became estancieros, while others had so many problems that they had to give up farming altogether.

Sheep farms prospered from the 1850s well into the 1880s, but the transformations suffered by the agrarian structure of the area in the last decade of the century must have greatly affected the fate of these farms. Decreasing profitability of the pastoral industry, increasing interest in other productive activities, such as agriculture and cattle raising, and the rapid increase in land values, all contributed to induce alternative uses for the land which until then had been devoted to sheep raising. Farmers had to choose either to become themselves *chacareros* or to rent out or sell their plots.[17]

Three Farmers in the 1860s

The story of three farmers who were settled in the province of Buenos Aires by mid-century will help to illustrate the different paths that could be followed by this type of rural family, and their situation in the decade of boom and crisis of the 1860s.[18]

(*a*) Michael Smith arrived in Buenos Aires from Ireland in 1854, when he was twenty-seven years old. He found work in Navarro, in pastoral establishments, and toiled his way up until he bought around 2,000 hectares by the Salado River. He had married Anne Lawler (also Irish, and illiterate), but they had no children, so he had to hire a fellow Irishman to help him on the farm. When he died in 1867, his farm was valued at 10,614 golden pesos, 8,400 of which corresponded to the value of the land. He had 2 flocks of mestiza sheep numbering 2,700 head, and a purebred Negretti ram, plus 4 milk cows and 14 horses and mares. All this stock represented approximately 17% of the capital invested. The main house was a two-room mud hut with thatched roof, containing a table, a cupboard, a few benches, and a bed. Two other huts and a kitchen also in mud and thatch were all that was to be found by way of buildings. A few trees and three folds completed the picture.

The lifestyle of Mr. and Mrs. Smith must have been quite modest. They sold their annual wool harvest to Michael Duggan, a well-known wool broker and estanciero, also of Irish origin, who not only commercialized Smith's produce, but also gave him credit. At Smith's death, he owed a total of 7,680 golden pesos to Duggan and to seven other people—among them his puestero. Obviously, the situation of this farmer was far from flourishing, yet the property was left undivided to his wife.

(b) Tomas McGuire was also an Irish farmer. He was born in 1815 and after migrating to Argentina had married Mary Kenny (also illiterate) in Buenos Aires in 1844. At his death in 1865, he left his wife and six children. The eldest daughter, Catalina, had left the household to marry Cristobal Quirno, but the other five children—three boys, two girls—lived with their parents in the farm they had in the county of Mercedes.

In an area of approximately 940 hectares McGuire raised 5,000 sheep and kept 2 milking cows. The main house was a two-room mud hut with thatched roof, while two other similar huts were named as puestos and probably were built in the limits of the estate. Three wooden folds and a dilapidated (*en mal estado*) fence made of wire were also listed in the inventory, as were tools and furniture, which included five drinking cases (for the sheep), a horse cart, a scale, three old pine tables, a cupboard, three pine benches, and a desk. At the request of the family, kitchen utensils were not listed. Total value of the farm was estimated at 12,961 golden pesos, of which 58% corresponded to value of land, 40% to stock, and only 2% to physical assets.

This inventory was made two years after the death of Tomas McGuire, and it was stated that during that period income resulting from produce sold had been enough to cover the regular expenses of the farm plus part of the interest on the capital owed by McGuire's family. In 1865, these debts amounted to around 7,000 golden pesos, and included three main sums: 1,800 golden pesos to the Mercedes branch of the Banco Provincia; 2,000$oro to Terence Moore, well-known merchant and moneylender, and 2,800$oro to Juan Esnaola, the remaining sum of a mortgage loan of 8,000$oro given by Esnaola to McGuire in 1862 to repay in one year at a monthly interest rate of 1%—later increased to 2%. Minor debts included 200 golden pesos to Liborio Torroba Hnos, and 165 golden pesos to Higinio Larrea, both of whom owned stores in the town of Mercedes.

In order to pay their debts, the family had to sell the farm. It was an unfavorable moment for them to do so, but the crisis had put down the price of sheep so low that by just selling the stock they could not have satisfied the requirements of the creditors. In 1869 they sold their property for approximately 16,800 golden pesos to Tomas Ledwith, an Irish settler who had acted two years earlier—together with M. Carty—as official assessor of McGuire's property. By then, the debts had climbed up to 10,000$oro, including testamentary expenses, so that around 6,800$oro were left to divide among the heirs.

(c) Michael Murphy was a more successful farmer. He also had been born in Ireland, emigrating to the River Plate area when still young, and marrying, in Buenos Aires, Isabel Scully. Neither of them had any property then, but already in 1855 Murphy's name is listed in the *Registro de contribución directa* of the county of Lobos as a landowner. After his wife's death he married a second time, one Lucia Ryan, who brought no property into the partnership. Murphy had nine children—six girls, three boys—with his first wife, and all of them were born and christened in the countryside.

When he died in 1864, at fifty-six, Murphy owned around 1,100 hectares in Las Heras and another 1,200 hectares in Lobos, rented two puestos also in Lobos, and owned a total of 10,000 sheep, plus cattle and horses for domestic use. His headquarters were in his estate in Las Heras, where the main residence was built of brick with a flat roof (*azotea*) and was well furnished with five pine beds, two chests of drawers, two washstands, two mirrors, three cupboards, twelve mahogany chairs, two armchairs, a sofa, a piano, three tables, a desk, abundant kitchen furniture, several lamps, benches, and so on. Next to the house, a kitchen was built in brick, an orchard was planted, and a shed was used to store produce and shelter tools. Two carts were parked outside, next to the folds. Four mud huts were distributed as puestos, with their folds and trees in different parts of the property. The establishment in Lobos, although a bit larger in extension, was more primitive in its organization, with only four mud huts with their respective corrals, and two flocks of sheep, totaling 2,900 head.

Obviously, Murphy had prospered as a farmer, and although he had a large family which probably also included his nephew Edward, he must have resorted to hired labor or aparcería to keep up production in his establishment. We have no information about the organization of his enterprise, but it is clear enough from the available documentation that the farm was for the Murphys a stage on the way that led from the lowest echelons of the rural order to becoming estancieros.

Tenant Farming

I have already mentioned that renting land was a widespread practice among pastoralists in the province of Buenos Aires, although for many reasons it was considered preferable to own the land upon which the enterprise—whether estancia or farm—was established. I have also said that being a tenant was considered an intermediate step toward the condition of landowner, although this was far from being achieved by all. As a result, a large number of small establishments were in the hands of tenants who worked with their families in very much the same way as farmers.[19]

Some of these tenant farmers were little more than sharecroppers or hired shepherds. They rented a puesto in a sheep estancia, cared for their own flocks, and paid the landowner the rent in money. But not always were they so independent in commercial and financial matters. Sometimes they had to sell the produce through the estanciero and ask his financial assistance. Their way of life was no different from that of the rest of the shepherds, as they lived in the same type of hut and led the same kind of hard life centered in everyday work with the flocks.

In spite of these resemblances between tenant farmers and estancia laborers, there was one fundamental difference: tenants did not work for an estancia, and except in critical years or in particular cases of close dependence resulting from a high degree of indebtedness to the landowner, these tenant farmers were able to accumulate and improve their situation. Thus it was not infrequent to find tenant farmers independently established upon a tract of rented land, organizing their small enterprise without belonging to the "clientele" of any particular estanciero, and working to enlarge their property.

What we have said about expenses, income, and profits in the case of farmers is also valid for tenants, except that rent was an additional expense to be considered. Rents were not too high throughout the period, but in a country with an expanding frontier, in an area which witnessed significant changes in its agrarian structure, and in a period of rapid valorization of land, rents suffered sharp oscillations from year to year, and were also very different from place to place. This made the situation of tenants unstable and uncertain, but at the same time it opened up opportunities because while certain areas were being entirely dedicated to agriculture with rents rising abruptly, new territories were put into production offering lands at relatively low rents. Pastoral tenants followed these opportunities, and one has the impression that they were always on the move.[20]

Whether in the hands of owners or tenants, sheep farms sprang up throughout the province during the pastoral era. These family enterprises found innumerable ways of surviving and expanding, each farmer devising his own particular method of organization and development. In the preceding pages I have tried to portray the basic features all of them shared, but one could as easily describe the differences to be found among them, as each farmer followed what he considered the best way to achieve his objectives—the survival and expansion of his enterprise together with the gradual improvement of his family's status and living conditions.

Notes

1. See, for example, William MacCann, *Two Thousand Miles' Ride Through the Argentine Provinces* (2 vols., London, 1853), I, 156–158, 279–282, and passim; M. G. and E. T. Mulhall, *Handbook of the River Plate Republics* (Buenos Aires and London, 1869 and 1885), edn. of 1869, 15–18, and edn. of 1885, 270; Wilfrid Latham, *The States of the River Plate* (London,

1868), 212–213; Emile Daireaux, *Vida y costumbres en el Plata* (2 vols., Buenos Aires, 1888), II, 243–245.

2. For example, Prudencio Mendoza, *Historia de ganadería argentina* (Buenos Aires, 1928), 170–171; Horacio Giberti, *Historia económica de la ganadería argentina* (Buenos Aires, 1961), chap. 5; José C. Chiaramonte, *Nacionalismo y liberalismo económicos en Argentina, 1860–1880* (Buenos Aires, 1971), 32–33; Jonathan Brown, *A Socioeconomic History of Argentina, 1776–1860* (Cambridge, 1979), chap. 7; María Saenz Quesada, *Los estancieros* (Buenos Aires, 1980), chap. 6.

3. Among Argentine writers, see, for example, Aldo Ferrer, *La economía argentina* (Buenos Aires, 1963); Ezequiel Gallo and R. Cortés Conde, *La república conservadora* (Buenos Aires, 1972); Giberti, op. cit.; Jacinto Oddone, *La burguesía terrateniente argentina* (Buenos Aires, 1967). Non-argentine scholars have stressed the same point; see S. Stein and B. Stein, *La herencia colonial de América Latina* (Mexico City, 1970); M. Carmagnani, *Formación y crisis de un sistema feudal* (Mexico City, 1976); F. Chevalier, *América Latina de la independencia a nuestros días* (España, 1979).

4. Brown, op. cit., and R. Cortés Conde, *El progreso argentino, 1880–1914* (Buenos Aires, 1979).

5. Studies on the rural development of the pampean region in the twentieth century recognize this point, and there is increasing interest in the expansion of both estancias and chacras. See, for example, Guillermo Flichman, *La renta del suelo y el desarrollo agrario argentino* (Mexico City, 1977)

6. There is a vast literature on the subject. Besides the already classic works of Lenin, Kautsky, Chayanov, and Wolf, there are a large number of more recent articles, papers, and books on the peasantry, as well as on different forms of petty commodity production. The following were particularly useful to my work: Samir Amin and K. Vergopoulos, *La question paysanne et le capitalisme* (Paris, 1974); Roger Bartra, *Estructura agraria y clases sociales en Mexico* (Mexico, 1974); Judith Ennew et al., "'Peasantry' as an Economic Category," *Journal of Peasant Studies*, 4 (4) (July 1977), 295–320; T. Shanin, *Peasants and Peasant Society* (Middlesex, 1971). For the Argentine case, see E. Archetti and K. Stolen, *Explotación familiar y acumulación de capital en el campo argentino* (Buenos Aires, 1975), which offers an illuminating approach to the study of the farmers.

7. MacCann, op. cit., I; C. Chaubet, "Buenos Ayres et les provinces argentines," *Revue Contemporaine*, 29 (1856–57), 233–261 and 473–500; Latham, op. cit., 91 ff.; Mulhall, op. cit., edn. of 1869, 15–18; Thomas Hutchinson, *Buenos Aires y otras provincias argentinas* (Buenos Aires, 1945), 316 ff.; Herbert Gibson, "La evolución ganadera," in *Censo Agropecuario Nacional: La ganadería y la agricultura en 1908* (Buenos Aires, 1909); J. C.

Korol and H. Sabato, *Como fue la inmigración irlandesa a la Argentina* (Buenos Aires, 1981).

8. Sometimes, immigrants came with capital (made in rural or in urban activities at home) and invested in the countryside, acquiring land and establishing their own sheep farm (*The South American Journal*, 4 October 1884; Mulhall, op. cit., edn. of 1869, 15–18). Locally, I have found almost no reference to urban dwellers becoming farmers, although it is not infrequent to find doctors in provincial towns, acopiadores, local merchants, and the like, buying land and establishing small pastoral enterprises. Probably, most of them employed hired labor, but it is also possible that some of them actually became farmers (Banco de la Provincia de Buenos Aires, Sucursal Mercedes, *Libro Mayor de crédito*, for the years 1880 and 1885—Archivo del Banco de la Provincia de Buenos Aires).

9. Estimates for the 1860s can be made on the basis of the data on the structure of property available for 1864. At the height of the pastoral expansion, we may assume that most rural establishments between 500 and 1,750 hectares were devoted to sheep raising. According to the cadastral map of 1864, in the counties of the sample used in chapter 2, 15% of the land was held in plots between 500 and 1,750 hectares, by 40% of the landowners in the area. It is quite probable that a large proportion of these establishments were sheep farms, but undoubtedly few under 500 hectares or over 1,750 hectares could be run as farms. For 1890, the amount of land held in plots that size increased considerably: 30% of the land, in the hands of 41% of the landowners of the area. However, as agriculture and cattle raising were already expanding in the area by 1890, it is hard to reach any significant conclusions as to the proportion of small or medium-size plots devoted to sheep raising.

10. The description of farms and farmers is based on references found in the following sources and texts: Chaubet, op. cit.; Hutchinson, op. cit., 316 ff.; Mulhall, op. cit., edn. of 1869, 15 ff., edn. of 1885, 270 ff.; *The South American Journal*, 19 September and 4 October 1884; Thomas Murray, *The Story of the Irish in Argentina* (New York, 1919), particularly chaps. 13 and 14; Kathleen Nevin, *You'll Never Go Back* (Boston, 1946); Korol and Sabato, op. cit., chap. 5; *Sucesiones*, AGN, Miguel Murphy (6842), Thomas MacGuire (6852), and Miguel Smith (8225).

11. *The South American Journal*, 4 October 1884.

12. Domestic expenditures could vary greatly from farm to farm, but in all cases these families resorted to the market to acquire most of what they consumed in the household. These farms were not self-sufficient, and although the main foodstuff was mutton (from the farmer's flocks) and sometimes milk and vegetables were produced in the farm, other

items of daily use, such as flour, salt, tea, *yerba,* and spirits, had to be acquired in the country store. Also kitchen utensils and most pieces of furniture and clothing were bought in the market.

13. The cost of shearing has been estimated for different decades throughout the period under study (see Table 15). Farmers probably paid the same as estancieros to the shearing squads, a cost which almost doubled between 1845 and 1885.

14. H. Gibson, *The History and Present State of the Sheep Breeding Industry in the Argentine Republic* (Buenos Aires, 1893), 115 ff.

15. Sucesión Murphy.

16. Ibid.

17. Mulhall, op. cit., edns. of 1885 and 1892, when describing the counties of the province of Buenos Aires.

18. Sucesiones Murphy, McGuire, and Smith; AGN, Sala III, Libro de Contribución Directa, Campaña: Año 1855 (33-5-14).

19. These references to tenant farmers are based on information found in Mulhall, op. cit., edn. of 1869, 15 ff. and edn. of 1885, 207 ff.; Latham, op. cit., 212–213; Gibson, "La evolución ganadera," op. cit.; Murray, op. cit., chaps. 13 and 14; Eduardo Coghlan, *Los irlandeses* (Buenos Aires, 1970).

20. Mulhall, op. cit., edns. of 1885 and 1892, particularly when describing the different counties of the province of Buenos Aires.

6

Wool Trade and Commercial Networks

Mais la richesse capitale de Buenos Aires . . . ce sont ses
laines, que se disputent, surtout depuis quelques années, la
France, l'Angleterre et la Belgique, et qui sont de plus en
plus employées à Liège et à Verviers, aussi bien qu'à Reims
et à Elbeuf.
—F. Belly, "Le Rio de la Plata et la République Argentine:
Coup d'oeil économique, statistique et financier," 1874.

Whether grown by estancieros or by sheep farmers, almost all the
wool produced in the province of Buenos Aires was sent abroad
to be used as raw material in the European textile industry. Exports of Argentine wool grew from under ten thousand tons in
the early 1850s to well over one hundred thousand tons in the
1880s.

In a previous chapter, I have analyzed the overall increase in
the exports of pastoral goods, and the oscillations imposed upon
this by international trade cycles. I am now going to examine the
problems involved in the actual operation of this commerce, particularly referring to the market and its institutions. Although the
pastoral era was a major period of Argentine economic history,
the wool trade has received little attention from historians and
social scientists concerned with the country's past. Therefore, as
the subject is almost unstudied, it is important to make clear the
objectives of this chapter.

In the first place, I will examine the international market for
Argentine wool as it developed between the late 1840s and the
1880s. Although I will alternatively mention Buenos Aires produce
and River Plate wool (which includes Uruguayan produce), this
first part refers mainly to wool exported by the country as a whole,
most of which was grown in the province of Buenos Aires.

Secondly, I will describe the commercial networks that operated

193

in Buenos Aires—city and province—analyzing their evolution in the years under study. In the third place, the various operations involved in the marketing of produce will be described briefly, and their cost estimated for the different decades of the period. Finally, I will refer to the relationship established between wool growers and commercial networks.

The International Market

The second half of the nineteenth century saw an unprecedented rise in the international demand for wool. As France, and later Germany and Belgium, followed the steps of Great Britain by going through an industrial revolution in textiles, production of woolens expanded and demand for the raw material rose accordingly.[1] Traditionally, European countries themselves had produced a large proportion of their raw wool requirements, and what imports they needed they had taken from their neighbors on the Continent—Germany, Spain, Russia—or from a few Eastern countries. But the growing demand for wool—coupled with a transformation in the economic structure of the manufacturing countries which drew away resources from the rural areas or diverted pastoral lands to other purposes—led to an increasing interest in wools from areas newly opened to production.[2] Great Britain had initiated this movement by introducing wool from her colonies, but it was not until the late 1840s that the Australian continent became the main supplier of English manufacturers.[3]

It was in this context that the River Plate territory started producing and exporting wool in the late 1840s, first to the United States and Great Britain, and later to the Continent (see Table 19). Its presence would become increasingly important in the French, Belgian, and German markets, as we shall see.

The development of the French wool textile industry led France to become a serious competitor of Great Britain in the second half of the nineteenth century. The Departments of the North had been traditionally an area of textile production, where as early as the late fifteenth century we find the local seigneur of Roubaix giving his subjects the right to "draper de toute laine"[4] (to manufacture wool fabric). It was, however, in the nineteenth century that this handcraft developed at a rapid pace to reach, after the

1850s, the stage of a modern manufacture. At a time when the taste for long wools that had concentrated the interest of the industry for so long was giving way to a preference for soft all-wool worsteds, the French were well prepared for the change, as they were specialists in the short merino types provided by their own flocks.[5] The introduction of technical innovations—particularly mechanical combing—and the discovery of coal in the northern districts contributed to the development of wool manufactures in Roubaix-Tourcoing, which became one of the world's main yarn-producing centers. A little later, the impulse reached Normandy—Elbeuf, Louviers, Lisieux, and Vire—until then characterized by the presence of a large number of small artisan enterprises that worked mainly with short wools.[6]

In the 1860s, the industry was to receive its greatest impulse. The growth and consolidation of a national market was followed by an important drive toward the international scene, as France expanded her exports of woolens. The crisis in cotton brought about by the Civil War in the United States was to favor this type of textiles throughout the world, and the French manufacturers profited from such a situation.[7]

The opening of the French market to foreign wools was an additional factor in this development. Until mid-century, heavy tariffs on raw wool hindered any large-scale imports of that product, and most of the raw material used was supplied by local flocks. A reduction in the tariff from 30% to 20% *ad valorem* in 1836, and further drops in 1851 and 1856, had encouraged the introduction of wools bought in Liverpool, whose real provenance is not easy to detect. By 1860, the French government responded favorably to the pressure of manufacturers and merchants by abolishing altogether the tariff on wool imported on French vessels; from then on, the participation of foreign produce in the local industry would grow steadily.[8]

Although countries like Russia, Austria, England (probably colonial wools), Uruguay, Algeria, and others at one time or another would appear as suppliers of raw wool to France, it was Argentina that became the main single source throughout the second half of the nineteenth century. By 1885, Buenos Aires wools accounted for 40.5% of the total wool imports of France.[9]

In the northern districts, the first manufacturers to mention

Table 19

Argentine Wool Exports to Belgium, France, Great Britain, USA, and Germany, 1861–1890 (in tons)

Year	Belgium tons	Belgium % of total	France tons	France % of total	G. Britain tons	G. Britain % of total	USA tons	USA % of total	Germany tons	Germany % of total
1861[a]	(5,805)	(37)	(3,747)	(24)	(598)	(4)	(1,597)	(10)	—	—
1862[a]	(10,887)	(48)	(4,887)	(21)	(1,662)	(7)	(4,449)	(20)	—	—
1863[a]	(11,101)	(39)	(6,621)	(23)	(2,341)	(8)	(6,676)	(24)	—	—
1864[a]	(18,590)	(50)	(6,199)	(17)	(3,441)	(9)	(7,957)	(21)	—	—
1865	(24,201)	46	11,849	23	2,559	5	11,713	22	—	—
1866	27,030	52	13,104	25	1,524	3	8,970	17	—	—
1867	32,485	53	19,734	32	2,382	4	2,623	4	—	—
1868	36,737	60	18,328	30	3,360	6	1,483	2	30	.05
1869[a]	(32,671)	(54)	(23,768)	(39)	(2,869)	(5)	(1,015)	(2)	—	—
1870	36,394	56	19,263	29	4,759	7	2,042	3	590	1
1871	38,229	53	10,836	15	7,296	10	7,264	10	—	—
1872	49,733	54	21,672	23	6,664	7	5,124	6	1,861	2
1873	42,754	51	22,298	27	7,684	9	4,565	5	760	1
1874	50,025	62	15,925	20	4,899	6	3,167	4	1,736	2
1875[a]	(45,749)	(61)	(21,081)	(28)	(1,779)	(2)	(1,408)	(2)	(3,059)	(4)
1876	47,577	53	19,562	22	3,696	4	3,233	4	4,780	5
1877	52,015	53	26,484	27	2,934	3	2,608	3	3,012	3

Year									
1878	38,416	47	26,632	33	906	1	3,311	4	2,990	4
1879	45,653	50	28,165	31	632	1	3,813	4	3,940	4
1880	38,114	39	35,699	37	1,198	1	3,827	4	6,875	7
1881	40,366	39	37,410	36	2,473	2	2,577	2	11,044	11
1882	42,986	39	39,556	36	5,738	5	991	1	14,397	13
1883	34,681	29	56,565	48	3,756	3	3,823	3	14,966	13
1884	33,029	29	55,502	49	1,989	2	2,153	2	17,425	15
1885	34,342	27	59,150	46	1,987	2	4,240	3	24,331	19
1886	32,429	24	61,979	46	5,322	4	3,796	3	22,485	17
1887	23,193	21	51,277	47	1,697	2	4,001	4	22,688	21
1888	31,045	24	57,927	44	3,191	2	2,332	2	30,334	23
1889	30,579	22	61,244	43	4,796	3	5,106	3	34,296	24
1890	21,378	18	55,821	47	4,814	4	6,502	5	24,731	21

[a]States only the wool exported through Buenos Aires customs. Percentages for these years have been calculated on total wool exports through Buenos Aires customs. Refers to figures in parentheses.

Source: Official statistics on exports, published in a regular bulletin which changed its name several times throughout the period, as follows:

1864–1868   Registro estadístico de la República Argentina
1869   No publication available
1870   Estadísticas de las Aduanas de la República Argentina
1871–1874   Estadística general del comercio exterior de la República Argentina
1875–1879   Estadística de la República Argentina. Cuadro general del comercio exterior durante el año de . . .
1880–1881   Estadística del comercio exterior y de la navegación interior y exterior de la República Argentina, correspondiente al año . . .
1882–1890   Estadística del comercio y de la navegación de la República Argentina, correspondiente al año . . .

River Plate produce were Toulemonde-Destombe of Tourcoing in 1855, and Amadée Prouvost of Roubaix in 1858. As yet, however, these wools were imported through London and Liverpool, and it was not until the 1860s that Antwerp and Le Havre replaced the British ports as the main points of introduction and marketing of River Plate wools. At that time, most of the produce was sent on consignment to the European ports, where manufacturers made their purchases with the aid of brokers. In 1866, however, the house of Jules Desurmont et fils, of Tourcoing, sent a representative to Buenos Aires to buy the wool on the spot. Afterward, other manufacturers of Roubaix, Tourcoing, and Elbeuf were to follow the same path, thus introducing important changes in the commercial procedures of the times. Although different ways of importing wool subsisted, there was a tendency toward direct purchases by the manufacturers, who thus avoided the cost of intermediation. The marketing role of the Continental ports decreased, and a long-standing entrepôt like Antwerp lost its predominance as French supplier. A tax on wool imported through foreign ports further contributed to this process, as Le Havre and Dunkirk gradually concentrated most of the import trade for Roubaix-Tourcoing and Elbeuf.[10]

Meanwhile, in southern France the trade with Buenos Aires had also expanded as a result of the development of textile manufactures in the Department of Tarn. Towns like Mazamet and Castres were the seat of an artisan production of woolens which dated back to the Middle Ages, and they had seen their trade consolidate during the first half of the nineteenth century. By the 1850s, however, this growth was in peril of being reversed by several economic factors—some internal to the region and others pertaining to the country as a whole. It was then that the initiative of the house Cormouls-Houlès of importing sheepskins in order to use their wool changed the fate of Mazamet, a town which still today is known in France as "la capital du délainage." The local manufacturers gradually shifted to this activity, and Mazamet became the purveyor of wool for the surrounding area. Most of the sheepskins imported by Mazamet came from the River Plate area. As early as 1851 the first skins arrived from Buenos Aires through Bordeaux, and five years later the Maison Périe of Mazamet sent a representative to that city to open a local office. Other houses followed, and by the beginning of the 1870s, seven of them operated branches in the River Plate.[11]

Therefore, for most of the second half of the nineteenth century, France was the main consumer of Buenos Aires wools. Argentine statistics, however, point to Belgium as the chief importer until 1882, on account of the role of Antwerp as an entrepôt.[12] In effect, a large part of the wool introduced through that port was sold at public auction to buyers from different manufacturing towns and found its way to the textile districts of Germany and northern France. Yet not all the produce was thus re-exported, and Belgium itself became an important consumer of River Plate wools when its main manufacturing town of Verviers started in the 1860s to compete with Bradford and Roubaix in the production of worsteds and the spinning of wool. The town became a center for scouring, carding, and combing River Plate wools, and exporting the scoured wool and yarn to other manufacturing countries. The early development of technical devices to remove the burr from the Argentine produce had proved an important advantage vis-à-vis the French and British manufacturers, who had to use hand labor to free the wool from the seed, thus adding to the cost of their imports.[13]

Throughout the 1860s and 1870s, while Antwerp was the main buyer of Argentine wools, Buenos Aires became the chief supplier of Belgium, covering by 1881, 86.77% of the produce introduced through its largest port.[14] By the end of the 1870s, however, the pre-eminence of Antwerp would start to decline. We have already mentioned how French manufacturers shifted to direct imports, and how Le Havre and Dunkirk challenged the role of Antwerp as the main introducer of Argentine wools into Europe.

The German towns, however, continued for some time to receive River Plate produce through Belgium, as they had done since the early 1850s. I have not been able to establish how much of the wool from Buenos Aires ended up in German hands, but the amount probably was far from insignificant, as there is continuous reference in German sources to this produce.[15] It was not until the unification of Germany that this country appeared as an important buyer in Argentine statistics (see Table 19). This was probably because in the decades before 1880, most of the wool destined to Germany went in through Antwerp and was registered as going to Belgium. Anyhow, the expansion of the worsted manufacture in the aftermath of the 1879 tariff and unification under Bismarck, was followed by an increase in direct imports of wool through the ports of Hamburg and Bremen, thus reducing the role of Antwerp as a supplier of the German towns.

Although France and Belgium, and later Germany, became the main consumers of Argentine wool in the second half of the nineteenth century, the initial stimulus for the expansion of production of wool in the River Plate had come from another source. In the 1840s and early 1850s, when the drive toward sheep raising started in the province of Buenos Aires, the main prospective buyers of Argentine wool were Great Britain and the United States. In the case of Britain, River Plate produce represented only a small part of the total imports of wool coming from different parts of the world, but mainly from the colonies, and much of it was re-exported to other European countries. With the increase of direct purchases by the Continental countries through Belgian, French, and German ports, Britain later held only a secondary place among the buyers of Argentine wool, although it continued to import sheepskins from the Plate.[16]

In contrast with Great Britain, the United States used most of the Argentine wool for its expanding manufactures. Its imports increased throughout the first half of the 1860s, reaching 23.95% of the total Argentine exports of dirty wools in 1865 (see Table 19). By 1867, however, this export trade suffered a severe blow when the American Congress voted a heavy tariff on the introduction of dirty wools in order to protect the local producers. As a result of this measure, imports of Buenos Aires wool into the United States decreased sharply, and the country became only a marginal buyer of Argentine produce.[17]

Competition and Prices

In the international market, Argentine wools had to compete mainly with those from Australia and the Cape. We have seen how by mid-century the consuming countries produced most of the raw material they required, and how, as demand increased and the economic structure of those countries changed, they began to import foreign wools for their manufactures. In 1860, 78% of the wool was self-supplied, but this proportion had been reduced to 48% twenty-five years later.[18] The search for alternative sources of wool led manufacturers to import from different countries, but by 1860 it was clear that the areas that could provide large quantities of the best type of produce at the most favorable prices were by far Australia, the River Plate, and in third place, the Cape. Although figures are far from exact, after 1873 Australia seems to have kept a comfortable first place as supplier, slightly increasing its participation in the market, while the River Plate—though in first place between 1863 and 1873—remained afterwards a secondary source with a decreasing share of the market.[19]

The commercial networks for these two competing areas were, however, entirely different. The introduction and marketing of Australian wool was carried out chiefly by the British, and in periodical auctions that lasted for several weeks the colonial wools were acquired by buyers from all over the world who came to London or Liverpool for that purpose. From the late 1860s to the 1880s, only around 50% of those wools was consumed in England, while the rest found their way mainly to the Continent.[20] This intermediation of Britain not only elevated the cost of the raw material for Continental manufacturers, but left French and Belgian ports out of the profitable business of marketing a significant amount of wool. More than once, attempts were made by the Antwerp merchants to attract Australian produce to their market, especially in the late 1870s and the 1880s when River Plate wool was finding other ways of reaching European manufacturers. The results were rather poor, however, and in that period no Continental port successfully challenged the pre-eminence of London as importer of colonial wools.[21]

The wool from the River Plate followed a different path. It was introduced in Europe primarily via Antwerp and Le Havre, where part of it entered directly—as it came to the order of particular manufacturers—and the rest was marketed in the ports in public auctions or in direct transactions with prospective buyers.

Although the commercialization of Australian and Argentine wool offered little field for competition, apparently the same did not happen with consumption. Actually, wool from the Australian provinces and from Buenos Aires province differed so little in quality—i.e., fineness, length of staple, elasticity—that prospective customers for either of them could easily switch to the other. There *were* certain preferences, however, as the condition in which each of them reached the market was not the same; their qualities were similar but never equivalent, and prices varied accordingly. Manufacturers complained of the dirty state of River Plate wool, which was not only in the grease, but also covered with clover seeds and burrs. The high cost of labor in Buenos Aires as compared to Belgium, for example, made it relatively unprofitable to clean the wool on the spot, and thus it was exported in a very poor condition to Europe, where sometimes it was acquired in that state by the manufacturers, and in other cases it was cleaned and scoured before being sold to its final consumers. British industry was reluctant to import the produce and preferred the better condition of Australian wool, but Continental—particularly Belgian—buyers had developed cheap methods for removing the burrs from the wool and therefore had no problem in importing it from the River Plate.

Certain minor differences in the type of produce also influenced the choice: for example, it was thought that River Plate wool was "better adapted for weaving the soft woolen tissues most produced in France than the harder fabrics of which we have nearly a monopoly."[22] Finally, price was also taken into account, and if wool from Buenos Aires was quoted at the lowest prices of all in the European markets throughout the period, it had a lower yield and involved additional costs because of its poor condition.[23]

These preferences and choices tended to become customary. We find, then, that the potential field for competition between Australian and Argentine wools in fact was limited by the high degree of specialization reached not only by the different national industries but also, within them, by regional and individual enterprises. According to Barnard: "Belgian scouring, combing and carding was predominantly confined to South American wool. . . . Though considerable quantities of colonial wool were imported there, it was generally processed to the specific requirements of other European industries. German and even French woolen manufactures, while they did use Australian wool, relied more heavily on South African and South American wools respectively."[24] Even in Great Britain, at the end of the century, the

North Country used mainly South American wools, while the West preferred Australian produce.[25]

Apparently the effect of competition between these two sources of wool very seldom was felt in concrete or specific cases. It is quite interesting to see that if in Britain, France, or Belgium, there was a certain degree of concern regarding the convenience and possibilities of using alternative sources for the raw material, in the producing countries, except in periods of acute crisis, there is relatively little reference to the problem of competition and even less to the opening up of new markets or to the expansion of the old ones. It seems that rather than increasing their relative participation in the world market of wool, both Australia and Argentina preferred to keep their traditional customers and adapt their production to the demands of these customers. This dependence was by no means costless, however, as throughout the period demand depended entirely on the state of the industry, and this, in turn, was a consequence of a variety of factors which had little—if anything at all—to do with the situation in wool exporting countries. Supplies, on the contrary, in the first instance resulted from conditions existing in the producing areas, and very often exceeded the immediate requirements of the European consumers. When this happened, the response of the market was to lower prices. Only then did producers react by limiting supplies and generally reversing the situation in a few years. In chapter 1 we saw how periods of crisis followed those of expansion, and how one of the producing areas (Buenos Aires) responded to those fluctuations. Australia had its own way of coping with these situations,[26] but in any case we see that the problem of the excess of supplies or of diminishing demand[27] was recurrent throughout this period.

But the absence of active competition also had its advantages for the producing areas, as their participation in certain markets was steady and significant, and therefore their influence therein was by no means irrelevant. Such is the case of Buenos Aires wool in Belgium—particularly in Antwerp and Verviers—and in France— Le Havre, Dunkirk, Roubaix-Tourcoing, Elbeuf, Mazamet. The weight of that produce in those markets was such that the course of prices was influenced by River Plate supplies, and therefore by the conditions existing in the Argentine at that time.[28]

Such a close relationship between the conditions of production in Buenos Aires and prices in the international market had no precedent in hides, nor would it have a similar relationship in wheat, as in both these cases—in different periods—staple exports of Argentina had little influence on the determination of international prices at the time of their local supremacy.[29] In our case, it was not only that contracting or expanding supplies of wool from Argentina made prices in European markets go up or down, but that producers and exporters in Buenos Aires more than once retained their wool stocks in order to bring about a temporary swing in price.[30]

Actually, producers were affected by, and exerted their influence upon, local prices in Buenos Aires (see Table 20), but as we can see from Figures 6 and 7, these prices ran almost parallel to those in Antwerp and Le Havre between the late 1850s and 1880s. Figure 7 shows the pattern of evolution of prices in Buenos Aires. Cyclical movements may be detected, but their length is not regular enough to allow for plotting moving averages. Therefore, we have chosen to calculate the linear trend, in order to perceive the long-term tendency of prices. The resulting line shows a slight rise between 1855 and 1892, but within that period, the first fifteen years reveal sharp and long fluctuations around the trend, while the following decades present a pattern of more frequent but less steep oscillations. Eliminating long-term influences, the graph shows, roughly, four cycles: 1858–1869, 1869–1876, 1876–1885, and 1885–1892. A period of rising prices followed the depression of 1857–1858 and lasted until the early 1860s, followed by a prolonged decline which brought prices down to their lowest level of the period in 1869. Recovery in the first years of the 1870s was swift but short-lived, although the trough of 1876 was not as low as that of the previous decade. Prices then tended to go up until 1880, but the trend for this decade was decline, with a short-lived recuperation in 1886.

These fluctuations ran parallel to those of River Plate wool prices in Le Havre between 1858 and 1878, and in Antwerp quite evidently in the last decade of those twenty years. London quotations had a similar course until 1876, but they appear as very depressed throughout the late 1870s and the 1880s. The movement of wool prices in Buenos Aires also coincided with Argentine economic

Table 20
Average Prices of Wool in Buenos Aires, 1855–1890

Year	Buenos Aires[a] $o × 10 kg	Buenos Aires[b] $o × 10 kg
1855	2.56	
1856	2.85	
1857	3.93	
1858	2.30	
1859	2.90	
1860	3.38	
1861	3.43	
1862	—	
1863	3.09	
1864	3.14	
1865	3.05	
1866	3.07	
1867	2.65	
1868	2.59	2.08
1869	2.19	1.81
1870	2.38	2.11
1871	2.73	2.45
1872	3.56	3.42
1873	3.00	2.75
1874	3.23	2.81
1875	3.31	2.99
1876	2.68	2.32
1877	2.99	2.71
1878	3.57	2.65
1879	3.02	2.72
1880	3.80	3.40
1881	3.29	2.79
1882	3.50	3.24
1883	3.46	2.80
1884	3.45	2.73
1885	2.94	2.13
1886	3.25	2.95
1887	3.36	2.88
1888	2.78	2.45
1889	2.99	
1890	3.25	2.54

Sources:
[a]1855–1861 *La Tribuna*, quotations at the Mercado 11 Septiembre for "Lana sucia mestiza fina" (yearly averages are my calculation, based on monthly quotations).
1863–1890 Juan Alvarez, *Temas de Historia Económica Argentina* (2 vols., Buenos Aires, 1929), II, 201–207. Quotations at the Mercado Central de Frutos de la Provincia de Buenos Aires—Sur, for "Lana fina madre" (yearly averages as above).
[b]*Anales de la Sociedad Rural Argentina*, 1868–1882: "lana mestiza buena"; 1882–1890: "Lana buena" (yearly averages, as above).

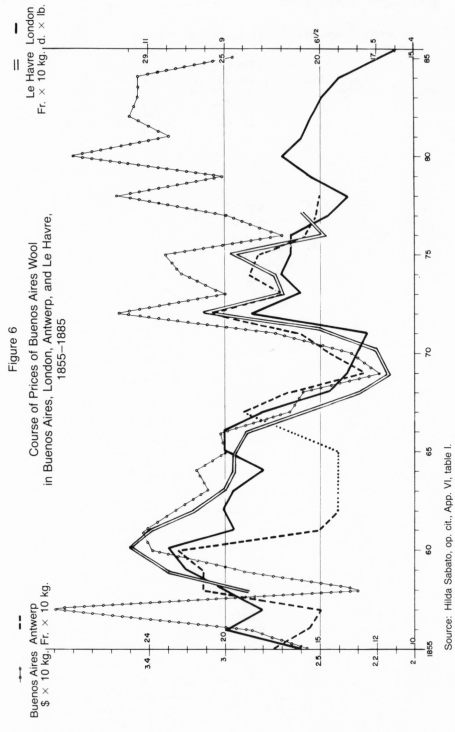

Figure 6

Course of Prices of Buenos Aires Wool
in Buenos Aires, London, Antwerp, and Le Havre,
1855–1885

Buenos Aires   Antwerp
$ × 10 kg.   Fr. × 10 kg.

Le Havre   London
Fr. × 10 kg.   d. × lb.

Source: Hilda Sabato, op. cit., App. VI, table I.

Figure 7

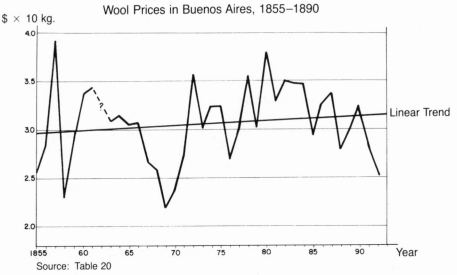

Wool Prices in Buenos Aires, 1855–1890

Source: Table 20

cycles, whose critical troughs generally occurred two or three years later than the corresponding ones in Europe.[31]

Fluctuations in wool prices resulted from the overall situation of the producing and consuming areas, and although international demand for wool was probably the main factor in the determination of those prices, the significance of River Plate produce in certain particular areas affected the course of prices in the corresponding markets. The expansion of the textile industry in the mid-century pushed the prices of wool to a high peak around 1860, which stimulated production in areas like Australia and the River Plate to a degree where supplies soon exceeded by far the requirements of the industry. Contraction of wool production followed and coincided in the early 1870s with the postwar boom on the Continent. Once more, however, expectations proved too optimistic and resulted in an excess of supplies in a situation in which the manufacturing sectors were suffering from the crisis of 1873–1876. A slight recovery in the late 1870s soon found a limit in the state of the industry, which was described in the early 1880s as follows:

Wool during the past year (1883) has followed an even course; there is nothing striking to record, and the results to both producers and consumers may be called fairly normal. Industrial profits being reduced to a minimum by the ever-growing competition, the tendency has been to seek compensation in the production of quantities. Hence a consumption of the raw material on a very large scale, but hence also a mass of yarns and goods in excess. . . . The trade rests on a basis both broad and sound, but it is stretched to the full and is, therefore, inelastic.[32]

This inelasticity reduced the margin of tolerance within the market, so that every increase in production, even if only moderate, resulted in a corresponding decline in prices.

The problem with prices, however, does not end in this short analysis of long-term trends and cyclical fluctuations in the different markets. Actually, we have greatly simplified the issue by choosing quotations for only one type of Buenos Aires wool in the local markets of the exporting country and of the main importing towns in Europe. In fact, Buenos Aires wool sold at very different prices according to its quality, condition and yield, and at times we find up to eight different quotations in the daily sheets published in the markets of the city of Buenos Aires, or in Antwerp, Le Havre, and London. We have further simplified the question by leaving aside shorter cycles, and—when using averages for each year—by avoiding the problem of seasonal, monthly, and even daily fluctuations. Finally, and this is perhaps more important in the context of this work, we have only mentioned prices paid *in the markets*, and these may reflect, but not necessarily coincide with, prices paid to the wool growers. In fact, it was very seldom that producers went to market to sell their wool in person, as more often than not they sold or consigned the produce to third parties who were in charge of selling and exporting it. It is to this question of intermediation between producers and the market that we shall now turn.

The Commercial Networks

Throughout the period under consideration, several commercial procedures coexisted and became standard in dealing in wool, as well as in sheepskins. Roughly speaking, the grower could either *sell* or *consign* his produce, but whichever he chose, he had more than one way of proceeding. He could *sell* the wool on the spot—either before or after the shearing—to a middleman who bought produce from different growers in order to resell it later in the market or to a particular customer. After collecting the wool, the middleman could proceed in turn to sell it or to consign it. If the latter, his steps followed the same path as those of the producer who chose to consign (see below). If the former, he had to find a customer locally—another middleman—or in Buenos Aires—*barraqueros*, exporters, agents of European houses. Sometimes, these middlemen acted by order of established clients in Buenos Aires.

From the late 1860s, and particularly in the 1870s and 1880s, producers could *sell* the wool on the spot to a new type of buyers: agents of European manufacturers who had to fill the orders of their houses and went to the countryside to acquire the produce, sometimes even buying the wool when still on the sheep's back.

Besides these transactions *sur place*, the grower also could choose to *sell* the wool in Buenos Aires. It was sent by cart, and later by rail, to the markets of Once de Septiembre and Constitución, where prospective customers attended to make their purchases. Representatives of export houses and agents of European firms were the main clients, although sometimes barraqueros and consignment brokers bought wool for speculation purposes or to fill orders of private customers.

If the producer chose to *consign* his wool, then he had to operate with a broker or consignee. Large and well-known estancieros sometimes consigned their produce directly to Europe,[33] but the majority of the wool growers sent it to Buenos Aires, where a number of established houses took care of the produce and sold it in the markets.

From this very brief description of the alternative ways the producer had to dispose of his wool, we perceive the complexity of the commercial networks that developed during the second half of the nineteenth century. We shall see, however, that most of these procedures were inherited from the past. Old networks were adapted to the new produce and had to expand in order to handle increasing amounts of wool and sheepskins, but until the late 1860s no new circuits had sprung up to compete with the established ones. Only when the French houses started to send their specialized buyers did an alternative procedure challenge the old ways, not to replace them, but to compete with them in the purchasing and exporting of wool. A more detailed analysis of the different links of the commercial networks will help us to understand how produce circulated and found its way to the export markets, and what was the actual position of the producers vis-à-vis those networks.

Export Houses. By mid-century, most of the trade between Argentina and the rest of the world was carried out by import-export houses that both introduced European goods and exported local produce either on order or on their own. The role of these houses, particularly those of British origin, has been already analyzed by Vera Reber, and other references to them have been made as well by other authors.[34] Therefore, I shall only mention their participation in the commercialization of the produce of sheep.

When wool became the staple export of the province of Buenos Aires, and later of the country as a whole, the pre-existing structure for commercialization was used for the purpose of dealing in the new goods. Wool was sent to Europe and the United States by those houses who took to export *frutos del país,* sometimes on order, at other times on consignment on their own account, and yet at other times acting as consignees for local growers. They used their contacts abroad to organize their operations, and very often had parent or associate houses in Europe with whom they worked in close relationship.[35]

During the late 1840s and the 1850s, most of the Argentine wool exported into Europe was introduced via Liverpool, as it was the British houses or those who worked in close association with them that monopolized the main part of the business. By the late 1850s, and more so in the 1860s, several French, Belgian, and German houses were operating in Buenos Aires, undertaking an increasing portion of the export of wool to the Continent, while Americans were in charge of the trade with the United States.

In Buenos Aires by 1850 Consul Hood identified fifty-six "British" mercantile houses (four of them in liquidation), twenty-one established by "citizens of France," and five by "citizens of the USA." Most of these houses had associate or parent houses abroad, although not infrequently they were headed by immigrants who had close ties with local merchants, thus becoming endeavors which used both native and foreign capital. Almost all the houses listed by Hood are described as general and/or commission merchants, suggesting little specialization.

A Belgian consular report of 1863 describes the different ways in which foreign trade was carried on in Buenos Aires, and emphasizes the role of merchant houses. The report lists houses of different origin, and explains which were the main products imported and/or exported by each. It identifies sixteen German, twenty-five British, four American, twelve French, four Swiss, and five Spanish and Argentine merchant houses. Finally, in 1886, a directory of Buenos Aires lists a total of sixty-nine export houses in the city, most of them still under foreign names.[36]

How did exporters acquire the produce? Part of the wool and sheepskins was bought or taken in consignment by these houses on the basis of personal contacts with the growers. Such was the case of large or well-known estancieros who avoided intermediation by dealing directly with exporters, not infrequently even before the shearing.[37] The main bulk of the wool, however, was bought to growers or brokers in the local markets. The produce from the countryside arrived in Buenos Aires at the marketplaces that had been built within the city limits and that acted both as domestic wholesale points and as centers for the processing of exports.

Until 1890, the Southern and Western Markets—later called Constitución and Once de Septiembre—were the most important points of arrival for wool and sheepskins. Before the age of railways, the produce was brought by cart and unloaded in the marketplace, where it was put for sale by a broker. From the late 1870s, most of the wool arrived by rail at the stations built next to the markets, and these were transformed to allow for the expansion of the trade. Warehousing facilities were always scarce, and most of the produce was transferred to private deposits or *barracas* within two or three days after arrival. The large sheds built in the 1880s within the old markets contributed to ease the situation, and in 1890 a huge market—the Mercado Central de Frutos—was inaugurated at the heart of the shipping district, and soon became the largest and most important place for the processing of exports. Exporters regularly attended these markets in

order to acquire the produce which was sold by private contract—
unlike what happened in other countries, public auctions never
developed into standard practice.[38]

By the end of the 1860s, the importance of import-export houses
in the wool trade had declined on account of two concurrent
factors: first, the rising demand for wool in the international market and the buoyant situation of the trade had led these houses
to put aside the practice of acting as consignment agents for local
growers, and to concentrate more and more on purchasing wool
on their own account, sending it to their European agents and
counterparts on consignment, and speculating with the difference
in prices, thus making large profits. The long crisis of the late
1860s, and the following setbacks in the early 1870s, however,
with their sharp oscillations in prices, greatly affected many of
those houses who had speculated in produce. And of those who
managed to stay in business, many chose less hazardous ways in
which to carry on the trade.[39]

A second factor in the decline of the importance of import-
export houses in the wool trade was the increasing role of agents
and representatives of European consumers in the Buenos Aires
market (see below).[40] Import-export houses, however, did not
completely disappear from the wool market, but there was an
increasing specialization as European merchants opened local
branches in order to acquire wool for export purposes. Firms like
Peltzer, for example, were established to deal almost exclusively
in the produce of sheep.[41]

Representatives of Foreign Houses. The high cost of intermediation, aggravated by an increasing tendency toward speculation
on the part of local import-export firms, led a few European houses
to try a system of direct purchases by sending representatives to
Buenos Aires with orders to buy wool or sheepskins. After the
pioneering experience of certain French textile firms in the late
1860s,[42] Belgian merchants and Continental manufacturers at large
saw the advantages of such a practice, and it became widespread
in the 1870s and 1880s.

During the shearing season, Buenos Aires was invaded by buyers from Europe who came to fulfill the orders of their employers.
They acquired wool by different means, sometimes making ar-

rangements with commercial houses for its shipping, but more often themselves taking care of all the operations involved in exporting the produce. Names like Schwarz (agent for C. Masurel of Roubaix), Donckier (acting for Juan Simonius and other firms of Verviers), and Peyras (sent by a house in Castres) are only a few examples in the early 1880s of these traveling buyers who very often represented more than one firm and not infrequently also did business on their own account.

The system was well described by the manager of the Buenos Aires office of the Bank of London and South America in a letter to his head office:

As you are aware, any wool orders that formerly came out to Buenos Aires came to the established business houses, but of late years we have seen practical men sent over here to select the special qualities of wool that their constituents required. These men, to begin with, brought letters of recommendation to the houses, in whose name all their purchases were shipped, and by whom all their financial requirements were attended to. The man who selected the wools did not require to occupy himself with anything more—it was the work of the house to whom he was addressed to provide the money each Saturday to pay for the purchases made, to pass the exchange, secure freight, render invoices and carry the business through to the end. The commission charged by the houses varied according to their standing, but competition on this side, and on the other side the desire to reduce charges as much as possible had brought this item to as low as 1%. . . . Presumably under the pressure of hard times the idea of saving this item altogether had induced some of the more important manufacturing houses to try to dispense with the services of the houses here. Then we saw the former sending over their own special agents to work here in the name of his house, Masurel, for instance. Others, who in first presenting themselves here, did so as the representatives of special houses, have gradually extended their relations and now execute orders for anyone. Thus a new class this spring are giving their attention principally to wool and sheepskins— pretending to buy safely on orders, but occupying a place in this market as exporters and drawers of exchange. As a rule they individually have no responsibility, they remain here during the shipping season and at its close go to Europe to look for orders for the following season bringing back with them such credits as they can get.[43]

Houses that made extensive purchases in the River Plate sometimes chose to have permanent representatives stationed in Buenos Aires, and a few of them established a local branch to carry out the operations.[44] In certain cases, they even proceeded to build their own warehouses to cater for their needs.[45] Once this organization was achieved, these houses could take advantage of their networks and facilities, and accept orders from other firms in Europe to buy and export produce on commission. In fact, some of these local offices originally created mostly by textile manufacturers to avoid intermediation became in themselves export enterprises of a new sort that performed many of the functions of the old import-export houses, not only in regard to their own purchases but acting as well on behalf of other Continental firms.[46]

The relationship between the local agents and their employers could have different degrees of independence. In some cases, the European houses chose to send one of their employees to manage business in Buenos Aires under strict orders from a home office. These agents were generally paid a regular wage plus a certain commission on purchases and profits. In other instances, manufacturers preferred to ask one or more professional buyers to purchase wool on their behalf, under contracts that specified the conditions of the relationship. These agents collected orders from different houses, and if they were established in Buenos Aires, they frequently became the regular representatives of more than one firm.[47]

All and every one of the existing ways of buying wool were practiced by these men. As Delpech describes it:

My father . . . purchased produce [wool and sheepskins] not only in the markets of Once and Constitución, but also—in large scale— in the warehouses that surrounded both markets, as well as in the important ones located in Barracas, like those of José León Ocampo, Balcarce, Rivas and many others, acquiring produce even in the countryside, in estancias, in provincial towns where local warehouses stored *frutos del país*.[48]

These foreign buyers would travel to the countryside by rail and on horseback in order to inspect lots of wool in faraway estancias and country stores, buying the wool and sheepskins from growers, middlemen, and barraqueros alike. After the purchase, they were in charge of sending the produce by cart or rail to Buenos Aires, to be baled and processed for export. Their presence contributed to change in more than one way the commercial practices of the early decades of wool growing, particularly in the late 1870s and 1880s.

We have already mentioned how the system of direct purchases from the manufacturing centers affected the pre-eminence of Antwerp and the role of its market in the international wool trade. In Argentina, the main consequences of this system were threefold, and all seemingly tended to improve the position of the wool grower vis-à-vis the commercial networks. In effect, the large influx of buyers to the local markets, plus the fact that most of them were ready to pay better prices than export houses because they saved on commissions and other expenses, pushed prices up in Buenos Aires to the benefit of growers and consignees. The extended practice of traveling to the countryside to make personal purchases in estancias and farms probably increased the opportunities for growers who previously had necessarily to rely on middlemen and consignees for the sale of their produce.

The favorable consequences of this practice should not be exaggerated, however, as in many cases purchases were made to local middlemen or in the markets of Buenos Aires through the regular channels. In any case, probably it was the large estancieros who benefited most from the possibility of dealing directly with the buyers, sometimes even selling the wool in advance of the shearing, upon arrival of these agents from Europe early in the season and at high prices.

The share that these agents of Continental houses had in the wool trade is very hard to estimate, but an indicator of the extent of their participation may be given by the fact that in 1886 only one of those agents—albeit a very important one: M. Gustave Heman, who worked for the house of Lorthois Frères of Tourcoing—was responsible for exporting over 30% of the amount sent to France or around 14% of the total wool exported that year from Argentina.[49]

Consignment Agents and Brokers. Whether shipped by import-export houses or by representatives of Continental manufacturers, most of the wool and sheepskins that arrived in Buenos Aires to be sold in the markets were sent from the countryside on consignment to local brokerage firms. The broker had to take care of the produce upon arrival, prepare it for sale, and find a customer. He charged a commission for his services, which varied in each case, but generally oscillated between 2% and 4% on sales and between 1/2% and 1% brokerage. The system was not new nor was it exclusive to wool or sheepskins, but in the period under study these became staple products of the countryside, so that most brokers undertook to deal in them, and more than a few consignment agents became quite specialized.

The main scene for the activity of brokers who dealt in the produce of sheep was the markets of Constitución and Once de Septiembre. It was there that they received the wool and sheepskins that came in carts and by rail to their order, that they prepared the lots for exporters and foreign buyers to inspect, and that they concluded most of their sales. The customers varied from day to day, but each broker had of course a number of regular clients.[50]

The business of brokers sometimes exceeded that of accepting consignment of produce. Very often they themselves became wool buyers by acquiring wool on behalf of a client, or making purchases on their own account, generally for speculation purposes. Frequently, large brokerage firms also entered the business of warehousing, providing services both to wool growers and buyers.[51]

The produce for sale was provided to consignment houses both by estancieros and by middlemen. Large wool growers generally had one preferred broker to whom they sent most of their produce every season.[52] At the same time, some brokers managed to attract a large number of small producers who consigned the wool directly to them, avoiding the intermediation of middlemen.[53] Most of the growers, however, had to resort to these, who in turn sent the wool on consignment to the Buenos Aires markets.

"Brokers form an important part of the commercial community of Buenos Aires," observed the *South American Journal* in 1885,[54] and apparently this had also been true thirty years earlier. As early as 1854, a group of brokers who practiced in the Mercado Once de Septiembre had founded an association called *Sala de Comercio de Frutos del País*. Among those who signed the minutes of the first meeting, we find the names of Saturnino Unzué, Juan Robbio, and several others whose houses were among the best known and more important consignment firms in Buenos Aires during the second half of the nineteenth century.

Going through the names of consignment houses which operated in Buenos Aires during the period under study,[55] it is apparent that although there were a large number of houses engaged in brokerage, few of them had a long life. The business probably appealed to many because it seemed to require little skill or capital: it was apparently a question of receiving produce and being able to find a customer who would buy it.

Furthermore, the business was almost entirely in the hands of natives or immigrants who were well established in the local society. But those who went into brokerage soon found out that if it was relatively simple to start such an enterprise, it was not so easy to keep it going. Successful performance was associated with both capital and connections. Capital was needed in the first place to advance cash to wool growers and middlemen—a common practice of the time which helped attract prospective consigners and to tie them to the broker. In a country where financial institutions were still in the process of organization, the commercial structure provided many of the services later to be in the sphere of banks.

On the other hand, connections were essential, because consignees operated in the market as links between exporters and producers, and their ability to appeal to both depended very much on the commercial as well as the social contacts the brokers were capable of establishing, not only in both those worlds, but also in the financial circles of the capital.

Other features may be associated with success. The expansion of activities within the trade and the diversification of investments were perhaps the most salient ones. In effect, brokers like Unzué and Duggan—to mention two of the more successful—were also dealers and speculators in produce. This was a risky business because if the periodical crises and the ups and downs in the prices of River Plate produce in the local and international markets provided the basis for large extra profits, they could also become the source of bankruptcy as well. Therefore, those whose fortune did not rely entirely on the one activity had more chance of coping with such crises and oscillations. And if we analyze the business record of Unzué and Duggan we find that they were not only— or mainly—brokers and dealers but also wool growers and land-owners, and that they invested in other fields as well.[56]

Barraqueros. The original role of the large number of barracas established in Buenos Aires and in provincial towns was to pro-vide storage facilities for the produce that was in the process of being exported. Hides, wool, and sheepskins arrived in the city and even if they went to the marketplace on the first day, they had to be stored until shipping, and had to be prepared—packed, baled—for export. Throughout the nineteenth century, these bar-racas provided warehousing facilities to exporters, consignees and producers, and they processed all the produce that went over-seas.[57]

The business was not an independent branch of the trade in wool or other goods, however, and if there were barraqueros whose first or main activity was to own and administer one or more warehouses, others were obviously better known as ex-porters or brokers. I have mentioned earlier how quite a few import-export houses, agents of foreign firms and consignment brokers built and operated their own barracas, not only to cater for their needs, but also to provide services to others as well. On the other hand, proper barraqueros never limited their business to managing their own warehouses, but also frequently acted as brokers to buy or sell, or speculated in produce, buying on their own account to sell later at a profit.

From the available information we see that the number of ware-houses in Buenos Aires tripled between the late 1850s and 1880s, while the amount of wool exported increased eightfold during the same period.[58] This bespeaks an expanding average capacity per warehouse, but it also has to do with the fact that large dep-ositories were built by the railway in the terminal stations annexed to the markets of Constitución and Once de Septiembre, where part of the wool came to be stored. By the late 1880s and early 1890s, the number of warehouses dropped, as the increasing share that these markets had in the storage of wool was further increased by the opening of the Mercado Central de Frutos, which soon concentrated most of the produce for sale and export, and offered ample and cheap storage facilities.[59]

No systematic information is available on the share that each warehouse had in the amount of produce processed for export. We have found only one estimate for 1866 of the number of bales of wool shipped by the main barracas of Buenos Aires during the season, where it is interesting to see that warehouses closely associated to exporters were those that had the biggest share in the business.[60]

In provincial towns, barracas became particularly useful in the age of railways, as the towns that housed a station soon developed into points of concentration for the wool of the surrounding areas. The storage facilities provided by the railways were far from sufficient to meet demand during the shearing season, and therefore warehouses supplied complementary services. Their numbers greatly increased in the early 1880s, especially in towns like Mercedes and Chivilcoy, where part of the wool was packed or baled in the local warehouses before being sent by train to Buenos Aires.[61] As in the case of the capital, in provincial towns barraqueros frequently acted as brokers and dealers in produce on a local scale.

Middlemen. Although large estancieros generally sent their wool directly to consignment or export houses in Buenos Aires, most wool growers sold their produce to middlemen, who collected it in different ways, and in turn proceeded to sell it or consign it. Legally, two types of licenses explicitly authorized the collection of produce in the countryside, those of *acopiador de frutos*, and of *mercachifle*. Actually, however, the trade was practiced as well by local warehousemen and wool growers, by country storekeepers and consignment brokers, and by all those who went to the countryside to make direct purchases from estancias and farms. Except for the already mentioned representatives of foreign houses, I shall include the rest in the broad category of middlemen, to mean those who collected wool and sheepskins to send to Buenos Aires for sale.

Throughout the province of Buenos Aires during the second half of the nineteenth century a large number of middlemen plied their trade, but because of their varied background and the differing scope of their operations, it is very hard to give an accurate picture of the business. Small purchases, like those made in 1852 by a Don Cruz Giles and a Pedro Arana with several growers and amounting to a few hundred arrobas of wool and skins, were contemporary to large transactions like those of Luis Aramburu, who besides collecting produce was an estanciero and a well-known broker.[62] Thirty years later, Delpech tells us how he bought wool from a local warehouse in Dolores, as well as from large estancieros who had acquired it previously from sharecroppers, tenants, and neighboring farmers.[63]

The collection of produce in the countryside had been practiced since the early decades of the century, when hides, bones, ostrich feathers, and so on were bought by storekeepers and *acopiadores*, sometimes from individuals who had "harvested" such produce from neighboring lands, at other times from Indians who came with the booty of their raids. Since the late 1840s, besides the produce of hunting and collection that they continued to buy, these middlemen started to purchase wool grown in the area, chiefly from small and medium-sized producers. Sharecroppers, tenants, and farmers who produced on a relatively small scale had no direct access to the market and thus had to rely on middlemen to sell their produce.

In the first place, we find the local storekeepers, or *pulperos*, who "are paid generally in produce—delivered during certain months of the year—by those who keep current accounts for the acquisition of foodstuffs and by establishments who receive cash advancements to meet their current expenses."[64]

In the second place, we may mention acopiadores, who traveled through the countryside acquiring produce from small growers. Sometimes these buyers made their purchase on behalf of Buenos Aires consignment brokers or export houses, but at other times they worked on their own account, bringing their produce to the capital to be sold in the markets or to warehouses or exporters.[65]

Thirdly, local warehouses in country towns concentrated the wool of the surrounding area, offering shelter to the produce that did not find a place in the railway shed. Very often, barraqueros became middlemen by buying lots from those who were in need of realizing the wool on the spot.[66]

Finally, large estancieros played the role of dealers in produce when they purchased wool (skins and others) from their own tenants and sharecroppers, as well as from neighboring farmers. In the case of the former, whatever independence aparceros may have had to dispose of their part of the wool in the 1850s and 1860s, it had disappeared almost completely by the 1880s, when all the produce went automatically into the hands of the employer, and the shepherd received his share in cash. As regards the latter, well-known estancieros like Bernardo de Irigoyen, Patricio Lynch, Senillosa, and others are known to have purchased wool from neighbors, sending it to their consignment agents or exporters.[67]

The relationship between grower and middleman was generally quite complex. Advances in cash, protection, and connections with the outside world were offered by these middlemen and had to be accepted by growers who, more often than not, had little real alternatives as to when and where they would sell their produce. Prices paid were generally low, and profits were high,[68] but there was a high degree of risk in operations that were primarily speculative in nature.

The Role of the Market

The functions of assembly and distribution of produce were performed in the market by the commercialization networks, whose main links we have described briefly above. These functions, however, involved more than just passing wool from one hand to another, for the produce had to be transported from the producing area to the markets, in the meantime being subject to a number of different operations. I shall now describe the main characteristics of the various activities involved in the marketing of wool, and the changes they underwent during the period under study.

Transportation. Wool was grown in faraway estancias and farms of the province of Buenos Aires and was consumed in the manufacturing centers of Europe. Many miles of land and water lay between the two and thus inland and oversea transport was involved. Throughout the period, the required means of transportation were provided by specialized agencies. From the early stages of the wool trade until the late 1860s, bullock carts were employed for the transportation of produce to the markets of Buenos Aires. These heavy two-wheeled carts could carry between 1¹/₂ and 2 tons of produce. They covered from 20 to 40 kilometers a day, depending on the weather conditions and on the state of the roads.[69] These roads were almost nonexistent, and when available, their state was far from satisfactory, particularly in wet months. Through more or less open country, the journey was slow and the wool often suffered from prolonged exposure to bad weather. Nevertheless, these carts provided a useful service, whose advantages would become a subject of public debate during the late 1860s and early 1870s, when their monopoloy of inland transportation was challenged by the railways.

Four rail lines were to make a dense railroad network in Buenos Aires in the following decades: the most important being the Western and the Southern Railways. The first section—10 km—of the Ferrocarril Oeste was inaugurated in 1857. In the 1860s it would reach Chivilcoy, in the heart of the sheep-raising district. With two branches—to Bragado and Pergamino in the 1880s—its extension was almost 350 km of railroad, a figure that would be tripled in the following decade. The first section of the Southern Railway reached Chascomús—114 km from Buenos Aires—in 1865. By 1880, different branches reached Las Flores and Bragado, a total of 560 km of railroad, also tripled by 1890. The other two lines—the Northern Railway and the Buenos Aires–Ensenada—totaled 130 km in the 1880s. Thus were the wool districts penetrated by this railroad network, which slowly but aggressively challenged the old carting system.[70]

Soon after the establishment of the first railway service to the countryside, the extensive advertising campaign carried out by different groups in favor of the new system of transportation was met by other voices who proclaimed the advantages of the old bullock carts. Some of the arguments then put forward are contained in the following two quotations, published by *The Brazil and River Plate Mail* and *Le Courier de la Plata* in 1865 and 1869 respectively:

This is the first year sheep farmers ever availed themselves of the railway to send their wool to market. Last year the Western Railroad was open to Lujan, one of the richest wool districts in the country, and yet scarcely an arrobe was sent by rail, but this year, although the season has hardly yet begun, more wool has come by rail than by bullock carts. This is as it should be. Aside from the very reduced freights which the railway charges, the conveniences which are offered completely shut out the bullock cart men from any competition. The loss which the poor sheep farmers suffer every year by the reductions made for damp wool amount to millions. A troop of carts six or seven days on the road and three days in the plaza, no matter how good the carts or careful the packing, rarely, very rarely, escapes without having some damp wool, and even supposing that there is none, the rickety character of the vehicle always affords a pretence for making the reduction. The railways, however, have put a stop to all this. Their waggons are waterproof, and the wool, if not sold within three days in the waggons, is housed in a magnificent depot, at, we believe, a most trifling cost; the farmer has nothing to fear from the rapacity of cartmen or extortion of comisarios. Last year many of our subscribers had to pay, after the three days had expired, one hundred dollars per day to the cartmen to keep their wool from being thrown out into the plaza. Then again, the farmer knows to a certainty the very day, nay, the very hour his wool arrives in the market, which is also an advantage not to be overlooked.[71]

And here is the argument of those who still preferred the cart:

Wool growers are concerned about the repeated handling suffered by their produce when unloaded into the waggons of the administration. Frequently, in most stations waggons are insufficient and in that case, wool suffers the following operations:
1. Unloading from carts into depots whether in Chascomus or other points.
2. Re-loading in waggons.
3. Unloading from waggons on arrival to depots.
4. Once sold, the wool has to be delivered to the buyer, and if by circumstances which are frequent enough it is not sold within three days of arrival and must be withdrawn from the depots and taken to other warehouses, it suffers further handling. . . .
Such is, sir, the reason of the preference accorded to transportation of wool by cart over its transportation by railway.[72]

But in the long run the railway won this uneven struggle, as its construction revolutionized transportation in more than one way. While by 1862 only 1,600 tons of wool arrived in Buenos Aires by railway, ten years later the amount had multiplied by ten, and by the 1880s almost all the produce went into the markets by train. In the end it had provided a faster and cheaper service than carts.[73]

Throughout the years, the service provided by railroads improved, with freight charges being lowered from an average of 0.09$oro per ton per km in the early 1870s, to 0.04–0.06$oro in the late 1880s.[74] Ten times slower and with a much smaller carrying capacity than trains (each railway wagon had four times the capacity of a cart), carts were also a more expensive means of transportation: in the 60s and 70s, freight rates were around 0.12 to 0.14 golden pesos per ton per km. The competition of the railway did not induce a reduction in those rates; on the contrary, in the 80s cart freights rose in absolute terms, probably because carts became a necessary complement to trains, in order to transport goods from the place of production to the railway stations. Journeys by cart thus became shorter but relatively more expensive.[75]

In response to the claims of railway customers, storage facilities were built, but they never filled the demand in the high season of the shearing. Voices of protest arose time and again pointing to this deficiency and to the damage it brought to the wool trade. Miles of railroad were built and stations were opened, not only in all the towns along the line but also in the middle of the plains, not infrequently in land donated by estancieros, who benefited from having the trains stop in their own back yard.[76]

Though information is scarce, I have estimated the cost of inland transportation for the three central decades of the pastoral era. Taking 100 km as the average distance traveled by the wool in the counties north of the Salado River, and presuming that the part of the journey made by cart diminished through the years, that cost dropped from 4% or 5% of the price of the wool as quoted in the Buenos Aires market to 3% or 4% in the 1870s, reaching a low of 2% or 3% in the 1880s.[77] The introduction of the railway had a great impact on rural production. Its more immediate effects were felt by the pastoral industry and by those who were in the business, as costs lowered, losses dropped, and profit margins rose. In the following decades, railroad construction was to expand the frontier of profitable activity in sheep raising both to the south and west, and at the same time, it opened the way to the successful expansion of agriculture throughout the province.

Overseas transport between Europe and South America was revolutionized around the middle of the century by the introduction of steam navigation in the route, which up to then had been covered by sailing vessels. Sixty to seventy days was the average length of the journey to Buenos Aires by sailing vessels, while it took them around one month to unload their cargo. Steamers shortened this time to one and a half months (thirty days to make the journey and around sixteen to unload), while the ratio of tonnage of the former to the latter was one to three.[78] Certain improvements in the port of Buenos Aires also contributed to shorten the time of loading, although it was only in the last decades of the century that a proper harbor was built.

The cost of overseas transportation varied greatly from year to year, and even from month to month, as freights were very much regulated by the variations in supply and demand, and in the case of freights for wool, the latter was entirely dependent on the state of the trade. Rates could double or even triple in the course of the same year, or they could sharply fall from one month to the next,[79] although most shipping lines charged a common rate.

In spite of these oscillations, we can clearly perceive a downward trend in the cost of freight for wool,[80] as the rates decreased from an average of 11 to 16 pesos oro per bundle of wool in the early 1850s to 3 to 5$oro in 1890.[81] This represented around 5% to 6% of the value of the produce in Buenos Aires in the 1860s, but only about 1% to 3% in the 1880s.

A large number of lines plied the route from Buenos Aires to different ports in Europe, but although Antwerp was the main port of arrival of Argentine wool, few of the steamers were Belgian. In point of fact British, French, and German lines were predominant in the period under study.[82]

Storage and Handling of Produce. Once the wool left the estancia, storage facilities were required at different stages of the trip to the hands of the consumer. Before the railway age, the produce generally went straight from the door of the farm or estancia to

Buenos Aires, where it was stored until shipping. We have mentioned that this service was provided by warehouses in the vicinity of the marketplaces, and since the late 1870s also by depots annexed to the markets. When railway stations became the gathering points for the wool of a certain area, sheds built by the railroads and local warehouses were both used for temporary storage before the wool was taken to Buenos Aires. In the final stage, when it arrived in Antwerp, Le Havre, London, or Bordeaux, wool had once more to be stored until sold.

A number of services were associated with warehousing. Baling the wool was perhaps the most important operation performed in the barracas, but there were other activities related to the handling of the produce which were repeated once and again, before and after storing or shipping. Such were, for example, those related to cartage and porterage, taring and making up, re-weighing, lotting and showing, sorting and grading, and the like. All these services were generally provided by agencies within the market, chiefly warehouses and brokerage firms, less frequently by export houses themselves. Sometimes, however, agencies that did not strictly belong to the commercial circuits, such as railway companies, covered certain aspects of these services—i.e., storage in sheds, handling by employees, loading, and unloading.

The cost of warehousing in Argentina and in Europe generally included a storage rent and additional charges for each one of the other services provided by the warehouse. In Buenos Aires, bills passed by barraqueros sometimes computed a commission, which apparently had to do with the brokerage functions they very often performed.

When the wool was sent to a consignment house in Buenos Aires or directly to an agent in the European markets, these were in charge of the produce and saw to it that all the services were duly performed and paid for. Afterwards, they included the cost in the bill sent to the consigner.

The total cost of warehousing and the associated services is hard to estimate, and it varied according to time, place, and particular circumstances (for example, the period of storage). A few references found in different sources speak of this cost as being, in Buenos Aires, around 3% to 6% of the price paid for the wool in the local market, and in Europe, from 2% to 4% of the same value.[83]

Market Information. Throughout the period under study this service was provided chiefly by wool-selling brokers and consignment agents. In the early days of the trade, information was made available in the markeplaces or at the local desks of export and consignment houses. With the creation of the Sala de Comercio de Frutos del País in the Once de Septiembre Market, and the Bolsa de Comercio, in the 1850s, the provision of this service improved, and periodical information on the state of the markets and the prices was sent to newspapers and journals to be published for the benefit of the public. In 1872, the *Revista de Frutos del País* started to appear regularly, including detailed reports and lists of prices.

In Europe, brokers, importers, and merchants had their own systems of publicity, and we find information on the situation of Buenos Aires wool in different markets in reports published by European firms such as Browne and Eagle, Ronald and Rodger, and others; in journals such as *L'Economiste Français, The Economist, Globus,* and others; and in various newspapers.

Risk-Bearing. Innumerable risks were involved in marketing the wool, but insurance was provided to cover only very few items. In the early days of the trade, *troperos* (bullock cart owners and drivers) bore the responsibility of safely reaching the Buenos Aires market, but when railways replaced the carts, this burden fell back on the growers.[84] Once in the capital, however, warehouses and brokers apparently were held responsible for the produce, but I have found no specific mention of insurance being taken to cover the risks, while, on the contrary, fire and marine insurance appears in every document referring to the shipping of wool and its storage abroad. The cost was between 1% and 2% of the insured value, and it was paid either by the grower or by the exporter, although in the first instance its negotation was generally in the hands of the brokers.[85] Insurance was provided by foreign companies that operated independently of the market. Before 1880, there were only two national insurance groups, but they depended upon British initiative, management, and capital.[86]

Provision of Credit. The provision of credit was one of the most important local functions performed by agents related to the market, but we shall leave its consideration to the next chapter, as it is an aspect of the more general problem of financial networks and procedures.

The Cost of Marketing

What was the cost of putting wool in the European markets? This is a simple question, to which I have no definite answer. We have seen how difficult it is to estimate the cost of the many services performed during the marketing of produce, and how this cost varied greatly with the time and circumstances. Nevertheless, I will attempt an approach to the problem which will arrive at tentative and approximate answers.

For this purpose, we must consider two stages in the transit of wool from the hands of the grower to the ultimate consumer. A first stage includes all the steps leading to the selling of the produce in the Buenos Aires market, and the second stage takes it from Buenos Aires to the Continental ports, where the ultimate buyer is found. As we have seen above, in many cases one of these stages is omitted when the grower sends the wool directly to Europe on consignment or when the buyer has an agent in Buenos Aires to purchase the produce. In these cases, however, most of the expenditures persist, except some of those related to commissions and brokerage.

Exclusive of these two items (eventual commissions and brokerage), the cost of sending the wool to Buenos Aires included cart and railway freights, and a tax on the transportation of produce called *guía de campaña*. Once in the marketplace, we may suppose that the wool was not sold immediately and therefore there was the cost of a few days of storage and of associated services to be paid for. In the early 1860s, all these would represent from 8% to 12% of the value of the wool already put in the market, but in the 1880s the cost probably oscillated between 6% and 9%.[87]

Once the wool was bought in Buenos Aires, it remained in storage until the time of its shipping. Carts then took the produce into the port, where it was loaded onto the ship to be transported to Europe, and once there, it was unloaded, and once more stored and prepared for selling. To the cost of warehousing and freight, we have to add, in this case, tariff paid to the customs in Buenos Aires,[88] and port dues. The cost of this stage of marketing was particularly influenced by the rate of freights and by the tariff imposed upon exports by the Argentine Government. In the early 1860s, wool sent abroad had to pay 5% ad valorem, but by the end of the decade this percentage went up to 8%, later to descend to 4% until 1874. The rate was 6% for the four following years, going up to 7% between 1877 and 1884, to decrease to around 3% up to 1888, when it was abolished altogether—only to be re-established three years later at 4%. These rates were paid ad valorem, according to the official value ascribed to wool, which more often than not did not reflect the real prices of the market. Therefore, these percentages are only an approximation of the actual incidence of tariff in the cost of wool.[89]

With these limitations in mind, we may estimate the cost of this second stage of marketing to be about 12% to 18% in the early 1860s, 8% to 16% in the 1870s, and 7% to 10% in the 1880s. To these expenses we have to add in most cases commissions and brokerage. Consignees both in Buenos Aires and Continental ports charged from 2% to 4% on sales and from 1/2% to 1% brokerage. In the case of foreign agents who purchased wool on behalf of European houses, their commission was generally 2% on the value of the wool bought in Argentina.

Finally, there were generally financial costs, including interest on money advanced by consignees and exporters to growers and middlemen, exchange brokerage, and banking commissions.

By adding up these percentages, which, as has been said, are calculated on the basis of the prices of wool in Buenos Aires (as per Table 21), we arrive at approximate figures.[90]

These costs were generally paid in each stage by those who were the temporary owners of the wool. Thus, when the grower consigned the produce directly to Europe, he was charged freight and other expenses (including brokerage and sales commission) by his agent, who discounted the cost from the payment of the wool he had sold at the prices on the European markets. In the opposite case, when a grower sold his produce at the door of his estancia, he did not have to cover any of the costs involved in the process of marketing, but he received a price that was nor-

Table 21
Cost of Marketing Buenos Aires Wool in Europe, 1860s–1880s
(as percentage of the price of wool in Buenos Aires)

	1860s	1870s	1880s
First stage[a]	8–12	7–10	6–9
Second stage[b]	12–18	8–16	7–10
Commissions	2–4	2–4	2–4
Brokerage	1/2–1	1/2–1	1/2–1
Totals	22 1/2–35	17 1/2–31	15 1/2–24

[a] "First stage" includes all steps leading to sale of produce in Buenos Aires market.
[b] "Second stage" includes costs of taking produce from Buenos Aires to continental ports.

mally lower than that which was quoted in the Buenos Aires markets.

We have no way of estimating the prices paid in estancias, which besides being regulated by the trends in Buenos Aires, probably depended very much on the particular situation of grower and buyer, and on their mutual relationship. For example, a large estanciero who sold his produce to a European agent probably realized much better prices than a small grower (tenant, farmer, or sharecropper) who sold it to a local storekeeper to whom he may have been indebted or to the owner of the land on which he raised his sheep. Other conditions, such as an eventual monopoly in a particular area or urgency on the part of the grower to dispose of the wool may also have influenced these prices.

When the grower consigned the produce to Buenos Aires, his agent charged him with the costs involved until the wool was sold at the prices quoted in the local markets.

The expenses related to the second stage of the marketing process were borne by the exporter or the European house if the goods were bought on commission. Only if the grower sold or consigned his wool directly to Europe did he have to cover these expenses.

In theory, the difference between prices paid for wool in the estancia and in Buenos Aires had to allow for expenses involved in the first stage of marketing, plus eventually the profit of middlemen, while that existing between local and European quotations should have covered the second stage, including profit of exporters. However, these differences proved on some occasions to be insufficient even to pay for marketing expenses, while on other occasions, they accounted for profits that exceeded by far any expectation.[91] Sharp oscillations in the prices actually paid for the wool—according to place, time, and circumstances—accounted for many of these situations, as well as for the increasing degree of speculation found in the trade. Consignment agents, brokers, warehousemen, and middlemen all were well aware of the possibilities offered by such a fluctuating market, and if the risk of purchasing produce on their own account to sell later at different prices was high, the extraordinary profits that could result from such an operation justified, in the eyes of many a middleman, such a risk.[92] The extent of these profits remains unknown to us, as we have no way of determining the prices paid and charged in each transaction.

Some Final Remarks

Since colonial times, the commercial role of Buenos Aires led to the organization of a merchant community in the city with connections both in the interior and abroad. After Independence, the opening up of the port to legal commerce with all the world, and of the city to the establishment of foreign merchants brought about a reorganization of the market. During the Rosas era, the expansion of cattle raising in the hinterland of Buenos Aires supplied the port with local produce for export, involving further changes in the organization of the market and its institutions.

Highly dependent on international demand, and very primitive in nature—almost no elaboration of the produce was involved—the export trade was organized on the basis of very flexible networks, easily adaptable to the different articles required by the market. Thus, except perhaps for jerked beef which had a different circuit, as later would frozen mutton and beef, all frutos del país were channeled through the same institutions. When wool appeared in the scene, no innovations were at first needed to commercialize it, and the old networks rapidly incorporated the article into their lot. The incredible expansion of the trade, and its relative importance as wool became the staple export, brought about a gradual specialization within the market. Thus, new agents exclusively dedicated to the wool trade made their appearance (agents of European manufacturers, wool exporters), while old agents accepted an increasingly specialized role (barraqueros who introduced the baling press in their warehouses, brokers who became wool brokers, etc.). This tendency, however, would soon find its limit. In effect, the sharp fluctuations of the trade due to its links with the international market proved decisive in bringing about diversification, and those who had initially entered the market to deal in wool had to extend the scope of their businesses. Meanwhile, old merchant firms had never altogether abandoned their other branches, and when the importance of wool declined in Buenos Aires, they promptly took their share of the new trades that promised expansion.

The relatively low degree of specialization required by the trade attracted many people, who tried their luck by associating themselves to certain links of the networks. We have mentioned the problems found by those who went into brokerage. Even more numerous were those attracted by the possibility of acting as middlemen, thus becoming acopiadores. The anonymous character of most operations carried out by them leaves us but little chance of estimating their successes or failures. However, the few references to the subject found in the sources all point out the risky nature of the business which, once more, favored those who could control more than one of the links in the trade networks, and who had diversified activities and investments.

The limits of these networks were never too clearly drawn, as the trade in wool became a profitable activity temporarily embraced by many, although controlled by few. However, this control did not develop into monopoly, as throughout the pastoral era different trade schemes coexisted or were successively adopted and dropped, giving way to a complex system where absolute control by few houses or individuals became ephemeral, if not almost impossible to attain. Nevertheless, some of them achieved an important place in the market, and contributed decisively to defining its shape and characteristics.

Among the most important names in the trade, both local and foreign houses are found. In fact, a first approach to the problem of the relationship between local and foreign capital in the trade would lead us to assert that while the commerce of wool in the first stage (estancia to port) was mainly in the hands of Argentines, overseas trade was predominantly carried out by houses with strong connections abroad and more often than not run by foreigners. A second reading would show us that this difference is not so clear-cut. Thus, agents of French and Belgian manufacturers going to the countryside to acquire wool were invading an area until then reserved to native consignees and acopiadores. At the same time, some of these foreign agents, as well as some of the export houses, developed strong ties with the local business world, and depended very little on foreign control or capital.

Nevertheless, contacts abroad were always essential to ensure good prices and advances of capital, which was so scarce in Buenos Aires. But even in this respect the situation was not so unfavorable to local interests. Thus, if until the 1850s British houses played a key role in River Plate overseas trade, in the following decades there was a diversification in the market while Belgian, French, and German firms appeared and gradually took over the wool trade. Competition then weakened the possibility of direct control. However, the market was far from being independent. In effect, overall dependence of the local wool trade on the conditions of the international market made the situation very vulnerable for the producing area. In this sense, and particularly in periods of declining demand, buyers abroad had all the advantage of being on the winning side, and were able to impose their conditions upon the local market.

The relationship of the growers vis-à-vis the commercial networks varied greatly with time and circumstances. Periods of increasing demand, such as the early 1860s and the late 1870s, were far more convenient for growers in general than were the years of crisis when overproduction dropped prices and middlemen and exporters could chose who to buy from and how much to pay. Although commercialization costs diminished greatly throughout the period, we know only of the prices paid in the markets, and therefore we cannot determine how much of this decrease reverted to the grower.

We have seen that different procedures for selling wool coexisted, and that a large number of buyers and brokers were attracted by the market. This should have favored growers in general. Not all producers were equally well placed to profit from such a situation, however, and possibilities varied greatly in each case. Actually, large estancieros were more independent, and therefore stronger, in relation to commercial networks than were small or medium-sized growers.

As we have seen above, well-known estancieros generally avoided the intermediation of acopiadores, and sold their wool directly to the exporter or sent it to a particular broker in Buenos Aires, or even in Europe. They had access to market information and very often could choose whom to sell to and when. Thus not only did they avoid costs of intermediation, but they could also realize better prices by negotiating personally with the buyers, sometimes retaining their wool while waiting for better quotations, or selling it in advance even before the shearing. Their situation was further improved by the fact that quite a few of the large estancieros themselves acted as middlemen, acquiring produce from tenants, sharecroppers, and neighboring farmers, and realizing profits on those wools as well as on their own. Finally, a small group of estancieros had a direct and relevant participation in the networks, acting as brokers and consignees of produce, managing well-established houses in Buenos Aires. Obviously, they could take advantage of their double role to obtain large benefits from both production and commerce.

Small and medium-sized growers, on the contrary, depended almost entirely on middlemen to dispose of their produce, although sometimes they were in direct contact with a particular broker or consignee. Their position in relation to both middlemen and brokers was generally far from favorable as they had to realize their wool right after the shearing in order to meet their obligations, and more often than not they were indebted to the same buyer, who had advanced credit to them for the season. Moreover, their access to information on market conditions, as well as their possibilities of choosing a buyer, were very limited. Nevertheless, in good times they seem to have shared the prosperity that followed the trade, particularly during the first decades of expansion.

Notes

1. In France, for example, consumption of raw wool rose from 83,400 tons in 1852 to 223,000 tons in 1885 (Paul Bairoch, *Revolución industrial y subdesarrollo* [Mexico City, 1967], 336).

2. Bairoch, op. cit., 337; Alan Barnard, *The Australian Wool Market 1840–1900* (Melbourne, 1958), 19–43; Claude Fohlen, *L'industrie textile au temps du Second Empire* (Paris, 1956), 364–66.

3. See tables on "Quantities of Sheep and Lambs' Wool Imported into the United Kingdom," published every year in *Parliamentary Papers, Accounts and Papers*.

4. Jacques Toulemonde, *Naissance d'une métropole: Histoire économique et sociale de Roubaix et Tourcoing au XIXe siècle* (Tourcoing, 1966), 14.

5. Barnard, op. cit., 24.

6. Fohlen, op. cit., 331.

7. Bairoch, op. cit., 337; Fohlen, op. cit., chap. 3.

8. While by 1840 foreign wools accounted for only 22% of the raw material consumed by French manufacture, by 1885 they covered almost 80% of that consumption (Bairoch, op. cit., 337).

9. Archives Nationales (Paris), *Douanes: Produits et dépouilles d'animaux (laines)*, 1890, F/12/6843.

10. See *Anales de la Sociedad Rural Argentina*, 19 (1885), 516, for a table on River Plate wools introduced through different European ports (1845–1883). See also Fohlen, op. cit., 132–37; Toulemonde, op. cit., 9–74; E. De Harven, *Le marché de laines à Anvers* (Antwerp, 1879); J. Blockhuys, "Le marché de laines à Anvers," *Rapports Commerciaux*, I (1881), 181–93; Félix Faure, *Le Havre en 1878* (Le Havre, 1879): *Bulletin de la Société de Géographie Commerciale de Bordeaux*, 1884 to 1890; Archives des Affaires Etrangères (France), *Correspondance Commerciale*, Buenos Aires (consular) vols. 3–15, 1838–1890.

11. Fohlen, op. cit., 331; G. Poulain, *La délainage et sa capitale Mazamet* (Castres/Mazamet, 1951).

12. Official figures quoted in Argentine sources as wool exports to Belgium differ from those found in Belgian statistics as wool imported from the Argentine Republic. Although small differences such as those found in the early 1860s could be accounted for by taking into consideration date of shipment as against date of arrival of produce, large ones like those of the 1870s and 1880s have to be otherwise explained. The role of Antwerp as entrepôt gives us the clue—probably Argentine and Belgian port authorities differed in the way they registered the wool that was to be re-exported.

13. Barnard, op. cit., 25–33; De Harven, op. cit.; Blockhuys, op. cit.; *Le Courier de la Plata*, 19 March 1868; *Recueil Consulaire* (R. Belgique) 12 (1866), 109–17; 16 (1870), 203–12; 18 (1872), 175; 20 (1874), 450–55; 21 (1875); Report by Mr. Grattan, British Consul at Antwerp, in "Abstract of Reports of the Trade of Various Countries and Places for the year 1859," *Parliamentary Papers, Accounts and Papers*, vol. LXIII, 1861, 4; Archives des Affaires Etrangères (France), *Correspondance Commerciale*, Anvers (consular), vol. 14, 1862–66, 233; *Parliamentary Papers, Commercial Reports* (b), Embassy and Legation, vol. LXIX, 1867, 305–7; L'Economiste Français, 1, No. 5 (1873), 132.

14. Blockhuys, op. cit., 182.

15. *Globus*, 20 (1871), 126; *Handels' Archiv*, 56, pt. 1 (1856), 281–82; 57, pt. 1 (1857), 616; 58, pt. 1 (1858), 697; Henkel, "Die Wolleproduktion der La Plata–Staaten," *Die Welthandel*, 2 (1870), 573–75.

16. "Report on Wool Imported to the United Kingdom," published every year in *Parliamentary Papers, Accounts and Papers;* Barnard, op. cit., 24; "Commercial History and Review," published yearly by *The Economist*, 1875 to 1890.

17. *The Merchant's Magazine*, 21 (1849), 558–59; 40 (1859), 345–46; J. C. Chiaramonte, *Nacionalismo y liberalismo económicos en Argentina, 1860–1880* (Buenos Aires, 1971), 64.

18. Barnard, op. cit., 39.

19. A. Sauerbeck, *Production and Consumption of Wool* (London, 1878); *The Economist*, Supplement: Commercial History and Review of 1872, 1876, 1885, 1890, and 1891.

20. G. Abbot, *The Pastoral Age. A Re-examination* (Melbourne and Sidney, 1971), 55–88; Barnard, op. cit., 25–33.

21. Blockhuys, op. cit.; De Harven, op. cit.

22. *The Economist*, 47 (20 July 1889), 927.

23. *The Brazil and River Plate Mail*, 6 April and 7 May 1867; 22 January and 7 October 1868; 8 January 1874. *Parliamentary Papers, Commercial Reports* (b), Embassy and Legation, vol. LXIX, 1867, 305–7.

24. Barnard, op. cit., 43.

25. Ibid., 43.

26. Abbot, op. cit., 58–88; N. G. Butlin, *Investment in Australian Economic Development, 1861–1900* (Cambridge, 1964), 60–61.

27. An example of these two different patterns of crises: Excess of supplies was the main cause leading to a significant drop in world wool prices in the late 1860s, while diminishing demand, resulting from a stagnant situation in British textile manufacture, led those prices to fall in the London market in the late 1870s.

28. Archives des Affaires Etrangères (France), *Correspondance Com-*

merciale, Anvers (consular), vol. 14, 1862–66, 396–97 (5 July 1866), and Buenos Aires (consular), vol. 10, 1873–78, 94–95 (30 January 1874); *The South American Journal*, 11 November 1880, 18; *The Economist*, Supplement on Commercial History and Review of 1886, 45 (19 February 1887), 27.

29. For tallow, see Tulio Halperin Donghi, "La expansión ganadera en la campaña de Buenos Aires," in T. Di Tella and T. Halperin Donghi, *Los fragmentos del poder* (Buenos Aires, 1968), 27; for wheat, Héctor Pérez Brignoli, *Agriculture capitaliste et commerce des grains en Argentine (1880–1955)* (Ph.D. diss., University of Paris, 1975), 171.

30. Such was the case in the early 1870s. After the Franco-Prussian War, international prices of wool rose suddenly in 1871–1872, and production in the River Plate area followed suit. The following season, when it was evident that the clip was going to be excessive, on account of a diminishing trend in demand, producers, middlemen, and exporters stocked wool in estancias and warehouses to avoid a drop in prices. In the short run they were successful, though they could not affect the long-term trend of prices. *L'Economiste Français*, I, No. 3 (1873), 73, and I, No. 15 (1873), 411; *The Brazil and River Plate Mail*, X (7 January and 8 March 1873); C. F. Woodgate, *Sheep and Cattle Farming in Buenos Aires* (London, 1876), 14. For a similar situation in 1881, see Blockhuys, op. cit.

31. According to Akerman, Juglar cycles in Europe during the second half of the nineteenth century had their critical years in 1857, 1866, 1873, 1882, 1890, and 1900 (Johan Akerman, *Estructuras y ciclos económicos* [Madrid, 1962]).

32. *The Economist*, Supplement on Commercial History and Review of 1883, 42 (23 February 1884), 25.

33. For example, in the early 1850s Senillosa consigned his wool directly to Kreglinger in Antwerp (*Archivo Senillosa*, AGN, Sala 7, 2-3-13, "Duplicado de factura" 2/8/52); in the late 1880s David Shennan made direct consignments to Liverpool and London (Herbert Gibson, *The History and Present State of the Sheep Breeding Industry in the Argentine Republic* [Buenos Aires, 1893], 211).

34. Vera Reber, *British Mercantile Houses in Buenos Aires, 1810–1880* (Ph.D. diss., University of Wisconsin, 1972) (although this dissertation was published in 1979, my citations are from the original text); Charles Jones, *British Financial Institutions in Argentina, 1860–1914* (Ph.D. diss., Cambridge, 1973); Jonathan Brown, *A Socioeconomic History of Argentina* (Cambridge, 1979).

35. Reber, op. cit., 149, 278; Jones, op. cit., 63; *Recueil Consulaire* (R. Belgique), 9 (1863), 238–41.

36. Archives of the Foreign Office, Public Record Office (London).

Argentina (F.O. 6), vol. 153; Consul Hood to Viscount Palmerston, Memorandum annexed to letter No. 10, Buenos Aires, 2 March 1850; *Recueil consulaire* (Royaume Belgique), 9 (1863), 238–41; *Gran Guía de la Ciudad de Buenos Aires, 1886* (Buenos Aires, 1886), 895.

37. For example, Bernardo de Irigoyen consigned his wool directly to Lumb (Jones, op. cit., 91–92); and Patricio Lynch to Zimmerman, Frazier, Co. (Brown, op. cit., 625).

38. Cf. Brown, op. cit., 105–8; Emile Daireaux, *Vida y costumbres en el Plata* (2 vols., Buenos Aires, 1881), II, 288–89 and 304–9; Estanislao Zeballos, *Descripción amena de la República Argentina* (3 vols., Buenos Aires, 1881/88), III, 236; Gibson, op. cit., 112–15; Carlos Lix-Klett, *Estudios sobre producción, comercio y finanzas de la República* (Buenos Aires, 1900), 268, 1257; Etienne de Rancourt, *Fazendas et estancias: Notes de voyage sur la République Argentine* (Paris, 1901), 251–53; De Harven, op. cit.; Blockhuys, op. cit.

39. De Harven, op. cit., 6–7.

40. Reber, op. cit., 278; Jones, op. cit., 63; BOLSA, Letter Book D 35-4, Buenos Aires Branch to Head Office, Confidential, 14 November 1879.

41. *Recueil Consulaire* (Royaume Belgique), 9 (1863), 238–41.

42. In 1866, the first of these agents from northern France, M. Adam, arrived in Buenos Aires, sent by the house Jules Desurmont et fils of Tourcoing (Fohlen, op. cit., 137–38; Toulemonde, op. cit., 49). He had been preceded in 1863 by Messieurs Daure, Guiraud, and Armengaud from Mazamet (southern France), who had opened a local office in Buenos Aires to sell woolens and buy sheepskins on account of M. Cormouls-Houlès.

43. BOLSA, Letter Book D 35-4, Buenos Aires Branch to Head Office. Confidential, 13 November 1879. Extensive reference to these wool buyers and the houses which sent them is found in the correspondence of BOLSA.

44. Such, for example, was the case with the houses of Cormouls-Houlès and Eugène Guiraud, of Mazamet, and Lorthois Frères, H. Caulliez, Masurel fils, Wenz et Cie., Wattine Bossut, of northern France.

45. For example: Barraca Armengaud, Barraca Del Pino (rented and later bought by E. Huc, associate to Guiraud); Bertram et Cie.; H. Caulliez; Duquennoy; Greffier fils; Soulas et Cie.; A. Trittan and others, all of which belonged to agents or associates of foreign houses.

46. The history of these houses and their agents in Argentina has yet to be written. Their business in Buenos Aires soon exceeded that of exporting produce of sheep for their own manufacturing concerns or on behalf of others. In the late 1880s, when speculation became the most profitable activity in the River Plate area, we find several of them in-

vesting in land, mortgage *cédulas*, government bonds, and shares of private and public companies (Emilio Delpech, *Una vida en la gran Argentina* [Buenos Aires, 1944], 133; Ets. Cormouls-Houlès, père et fils, Private archives; *Correspondence*, Livre Particulier, 1884–87 and 1887–91).

47. The correspondence of the House Cormouls-Houlès cited in note 46 abounds in references to these agents. Further references in Delpech, op. cit.

48. Delpech, op. cit., 63–64 (my translation).

49. *Anales de la Sociedad Rural Argentina*, XX (1886), 584.

50. Brown, op. cit., 105–8; Daireaux, op. cit., II, 304–9, 321; Zeballos, op. cit., III, 236–37; Gibson, op. cit., 170–78; Lix-Klett, op. cit., 268–69; *Tribunales de Comercio de Buenos Aires*, AGN, Legajo P/274 "D. J. Perez contra Dn. Ildefonso Torres y Gervasio Orozco," 1857.

51. For example Berrotarán (the family had a warehouse already in the 1850s and were still in the business in the 1890s); Unzué (same); Ber, Casey, Duggan hnos., in the 1880s; Otero y Cía., in the 1870s.

52. Such was the case, for example, of Felix Alzaga, who always consigned his wool to Saturnino Unzué (Delpech, op. cit., 110).

53. For example, Casey and Duggan used to attract the wool of the Irish small wool growers (J. C. Korol and H. Sabato, *Como fue la inmigración irlandesa a la Argentina* [Buenos Aires, 1981]).

54. *The South American Journal*, 22 (24 January 1885), 39.

55. In the 1880s I have detected almost 200 names of brokers and consignment agents working in Buenos Aires. Lists of existing consignment houses are available in the different *guías* published throughout the period (see Bibliography).

56. As mentioned in chapter 5, the Unzués were a traditional family in Buenos Aires. Francisco Unzué had arrived from Spain and occupied colonial posts during the eighteenth century. Second-generation Unzués became estancieros and merchants, one of them establishing the firm Saturnino Unzué e Hijo to manage their business and property. Michael Duggan, on the contrary, arrived in Buenos Aires around mid-century, with no fortune. He soon established contacts with the local Irish community, worked at first in the city, and then in the countryside as shepherd, soon becoming a sheep farmer. In the 1880s he possessed over 25,000 hectares in different partidos of the province, had a prosperous consignment house in Buenos Aires, and entered into partnership with Casey to invest in different concerns—not always successfully.

57. J. A. Víctor Martin de Moussy, *Description géographique et statistique de la Confédération Argentine* (Paris, 1860/73), 124–25; Benjamín Vicuña Mackenna, *La Argentina en el año 1855* (Buenos Aires, 1936), 124–26; William MacCann, *Two Thousand Miles' Ride through the Argentine Prov-*

inces: Being an Account of the Natural Products of the Country and the Habits of the People (2 vols., London, 1853), vol. 1, 216–17, Delpech, op. cit., 34–36, 64–69.

58. Information on the number of existing warehouses is to be found in the different volumes of the *Registro estadístico de Buenos Aires*. Lists of barracas are published in the guías cited in the Bibliography.

59. Zeballos, op. cit., II, 236; Gibson, op. cit., 173–78.

60. *The Brazil and River Plate Mail,* 7 July 1866, 296. Those barracas that appear with the largest share in the business, with over 6,000 bales (2,200 tons) in the season, are: Ch. Bove, J. C. Mohr y Cia., W. Bertram, Mohr and Clausen, J. Balcarce, John Smith, Prange and Co.

61. Data on the amount of wool sent to Buenos Aires from the different stations by the Western Railway is to be found in Ferrocarril del Oeste, *Memorias del Directorio* (Buenos Aires, 1880–1884).

62. *Tribunales de Comercio de Buenos Aires,* AGN, Legajo G/109, "Don Cruz Giles contra Don Pedro Arana and José González Molina contra Gil Cabrera," 1852–1854.

63. Delpech, op. cit.

64. *Anales de la Sociedad Rural Argentina,* XI (1877), 132 (my translation); see also Martin de Moussy, op. cit., 18.

65. *Anales de la Sociedad Rural Argentina,* XI (1877), 132–34; Zeballos, op. cit., III, 236–37.

66. Delpech, op. cit., 34–36.

67. Jones, op. cit., 63; Brown, op. cit., 117; Daireaux, op. cit., II, 321; *Archivo Senillosa,* AGN, Sala 7.

68. Zeballos, op. cit., III, 236–37.

69. Martin de Moussy, op. cit., 566–67.

70. Eduardo Zalduendo, *Libras y rieles* (Buenos Aires, 1975) chap. 6.

71. *The Brazil and River Plate Mail,* 22 December 1865, 656.

72. *Le Courier de la Plata,* 7 February 1867 (my translation). See also *The Brazil and River Plate Mail,* 22 November 1865, 606–7; Eduardo Olivera, "Nuestra industria rural en 1866," in *Miscelánea* (2 vols., Buenos Aires, 1910), vol. 1, 85–102.

73. AGN, Sala 10, 32-5-3: Policia. Tabladas, Corrales y Mercados, 1861–63; Ferrocarril del Oeste, *Memorias del Directorio,* 1872–83; William Rögind, *Historia del Ferrocarril del Sud, 1861–1936* (Buenos Aires, 1937).

74. Ferrocarril del Oeste, *Memorias del Directorio* (1873), 17–18; Gibson, op. cit., 185.

75. Ferrocarril del Oeste, *Memorias del Directorio* (1873), 17–18; Gibson, op. cit., 185; T. Hutchinson, *Buenos Aires y otras provincias argentinas (1860)* (Buenos Aires, 1945), 136; Olivera, op. cit., 85ff.

76. Ferrocarril del Oeste, *Memorias del Directorio* 1871 to 1883; William

Rögind, op. cit.; *Anales de la Sociedad Rural Argentina*, XVII (1883), 565; Gibson, op. cit., 178–82.

77. Taking into account the location of the railways and their extension in the different decades, estimates were made on the basis of considering that in the 1860s the whole journey was done by cart; in the 1870s half the journey was done by cart and the rest by train; and in the 1880s, 20 km were done by cart and 80 km by train. As we do not have estimates for the cost of production of wool, all the percentages in this chapter are calculated on the basis of the price of wool in the Buenos Aires market, as per Table 20, first column.

78. Reber, op. cit., 113–16.

79. For example, in 1888 freights by steamer to Antwerp were 15 francs per bale of wool in January, 10f. in July, and 38f. in December (*Registro estadístico de la República Argentina*, 1888).

80. Freights were generally calculated on the basis of the volume or the weight of the goods to be transported, but apparently wool was always rated by volume (José Hernández, *Instrucción del estanciero* [Buenos Aires, 1964], 265–67; *Parliamentary Papers, Commercial Reports*, [b] Embassy and Legation, vol. LXXIII, 1876, 204).

81. Reber, op. cit., 96–98; Faure, op. cit., 245; *Archivo Senillosa*, AGN, Sala 7, 2-6-13, Duplicado de factura, 2/8/52; Ets. Cormouls-Houlès, Private archives, *Correspondence*, Livre Particulier, 1872–1878; *Registro estadístico de la República Argentina* (1884–1890); *The South American Journal*, 18 (22 December 1881), 1–2, and vol. 22 (7 March 1885), 107.

82. *Recueil Consulaire* (Royaume Belgique), 15 (1868–1869), 137–46; Lix-Klett, op. cit., 616; *Registro estadística de la República Argentina*, 1879.

83. The calculations are based on evidence included in different sources: *Archivo Senillosa*, AGN, Sala 7, 2-6-13, Duplicado de factura, 2/8/52; *Tribunales de Comercio de Buenos Aires*, AGN, Legajo G/109, Gerding contra Coqueteaux, Sarrasin y Vignal, 1852–1854; MacCann, op. cit., 216–17; Faure, op. cit., 245; Gibson, op. cit., 185. The correspondence of the house Cormouls-Houlès, pere et fils, makes recurrent mention of these costs.

84. Olivera, op. cit., 85–102.

85. *Archivo Senillosa*, AGN, Sala 7, 2-6-13. Duplicado de factura, 2/8/52; Faure, op. cit., 245. The correspondence of BOLSA repeatedly refers to these matters.

86. See Reber, op. cit., 244–45; Jones, op. cit.

87. Calculations are based on evidence found in Ferrocarril del Oeste, *Memorias del Directorio*, 1871 to 1883; *Registro estadístico de la República Argentina*, 1884–1890; Gibson, op. cit., 185; Olivera, op. cit., 85ff.; MacCann, op. cit., 216–17; Thomas Hutchinson, op. cit., 136; *Archivo Senillosa*,

AGN, Sala 7, 2-6-13, Duplicado de factura, 2/8/52; *Tribunales de Comercio de Buenos Aires*, AGN, Legajo G/109, Gerding contra Coqueteaux, Sarrasin y Vignal, 1852–1854; Ets. Cormouls-Houlès, Private Archives, *Correspondence*, Livre Particulier, 1872–1878.

88. Throughout the period under study, dirty wool paid no introductory tariff in Belgium, Great Britain, and Germany. In France, as we have seen, such a tariff was abolished in 1860 for wool imported directly to French ports. The United States passed a very high tariff on the import of dirty wool in 1867, and the situation remained stable all through the rest of the century.

89. For a discussion of the relationship between the official prices of produce (on the basis of which *ad valorem* tariff was calculated) and the actual market prices, see R. Cortés Conde, T. Halperin Donghi, and H. G. de Torres, *Evolución del comercio exterior argentino . . . Exportaciones* (Buenos Aires, 1965) (mimeo). The differences between official prices of wool and actual ones is mentioned time and again by the sources. See, for example, Archives des Affaires Etrangères (France), *Correspondance commerciale*, Buenos Aires (consular), vol. 8, 1868–1869, fs. 352–53, 411–12, 10 July 1869, 12 August 1869; *Recueil Consulaire* (Royaume Belgique), 16 (1870), 210; *Anales de la Sociedad Rural Argentina*, III (1869), 222; F. Seeber, *Apuntes sobre la importancia económica y financiera de la República Argentina* (Buenos Aires, 1888), 132.

90. I do not have data on the cost of commercialization for other produce to be able to make a comparison, but Australian estimates on the cost of the second stage, plus brokerage and commissions, oscillated between 10% and 15% for the late 1870s and the 1880s, which is not too different from my figures. See Butlin, op. cit., passim.

91. For an example of the first of these situations, see, for 1869, Archives de Affairs Etrangères (France) *Correspondance Commerciale*, Buenos Aires (consular), vol. 8, 1868–1869, fs. 312–13, 10 May 1869; *Le Courier de la Plata*, 5 May 1869. For 1879, see *Le Courier de la Plata*, 20 September 1879. for the second case, see, for 1872, Archives de Affaires Etrangères (France) *Correspondance Commerciale*, Buenos Aires, vol. 9, 1870–1872, f. 364, 21 March 1872.

92. Delpech, op. cit.; Zeballos, op. cit., 236–37. In *Tribunales de Comercio de Buenos Aires*, AGN, Legajo G/109, 1852–1854, I found an interesting example of speculation. A middleman offered a small grower 35$ per arroba of his wool, and he stated he was able to sell it at 55$ per arroba.

Finance and Credit in the Pastoral Sector

Les indiens, les voleurs, la Banque. Nous n'accouplons pas
ces mots sans motifs. Les indiens profitent de la situation
que la guerre du Paraguay a faite au pays pour piller nos
frontières; les voleurs pour saccager nos campagnes; la
Banque pour vider nos poches.
—*Le Courier de la Plata*, 12 January 1866.

Monetary Problems in Argentina

"Throughout the nineteenth century, the history of the Argentine
currency had been a melancholy tale of over-issue of paper money
. . . interspersed with temporary attempts at stabilization, the
printing press being the favorite source of revenue."[1] These few
words by A. G. Ford underline one of the main features of the
Argentine economy: monetary instability. At the heart of the problem lay the relationship established between paper money issued
for current transactions—and which through most of our period
was not uniform throughout the country—and gold, used for
international transactions. As Ford has pointed out, "Herein lay
the problem of Argentine monetary management—to link the two
standards together at a fixed rate of exchange for this 'export'
economy."[2]

Gold was not produced in Argentina, and therefore specie reserve resulted only from a positive balance of payments and from
private sources. However, paper currency was generally issued
without even an attempt to preserve a sound relationship with
such a gold reserve; therefore giving way to recurrent crises in
the monetary field.

In the province of Buenos Aires, large issues of inconvertible
paper money, frequent and sharp oscillations in the value of gold,
and an almost continual devaluation of the paper currency were

characteristic of the 1850s and early 1860s as a result of the variable price of and demand for Argentine exports in the European market and the financial measures taken by the local government to face the problems brought about by internal struggles and civil war. The amount of money in circulation expanded, and except for the year 1857 when there was a shortage of cash during the export season, the decade was one of abundant and even overabundant paper money.[3] Speculation was the order of the day. As Hinchliff has observed:

> Under those circumstances, no wonder many respectable men do not dare to set foot on regular commerce . . . and see themselves reduced to speculation at the stock exchange. Gangs of petty brokers and intruders, little else than common gamblers, stick around the Bolsa.[4]

The government tried to stop this trend by implementing several financial measures and ceasing the issuance of paper currency. By 1864, devaluation had stopped and soon the trend reversed, leading to a process of revalorization of the paper peso and to a period of scarcity of currency. The slight decrease in the amount of currency in circulation by 1861, together with the boom in wool production and trade in the early 1860s, had contributed to this reversal, and soon Olivera would announce that the shortage of cash was hampering commerce and agriculture.[5]

In order to put a halt to this process of revaluation, the Sociedad Rural demanded government intervention, asking for the establishment of an official value for the paper peso and for its free convertibility at a fixed rate of exchange. In spite of the opposition to these measures voiced by certain newspapers like *La Nación Argentina* and *El Nacional*, the government created the Oficina de Cambio (Bureau of Exchange), which was authorized to exchange paper pesos into pesos fuertes at the rate of 25 to 1.[6]

These monetary problems were at the root of the crisis of 1867, which so much affected the pastoral interest in the province, though they were accompanied by other important causes too, such as overproduction of wool and the drop in prices and demand for this product in the international market.

When, by 1870, recovery occurred, it was the result of several favorable circumstances, but in the monetary field perhaps the most important ones were a better balance of trade brought about by the improvements in the prices of wool and the rise of exports in 1871 and 1872, and the entrance of Brazilian gold during the Paraguayan War.[7] Together with the receipt of increasing amounts of foreign capital by way of loans, these circumstances allowed for an expansion of the amount of currency in circulation, while the exchange rate was always under control, and paper pesos were freely convertible into gold. Specie reserves increased greatly. Banks were created and credit swelled.[8] But again expansion of credit and speculation went hand in hand. Foreign capital started to emigrate in view of the European crisis; the balance of trade became increasingly negative as a result of a drop in the price of Argentine produce abroad; and political problems combined to generate financial insecurity and a rush for gold. At this point crisis ensued.

By 1875, President Avellaneda explained the situation in this way:

> In the last few years, large amounts of money, originated in loans negotiated by the Nation and this province in London, flowed to Buenos Aires. Thus, its accumulation in banks, the low interest rates and the appealing facilities of credit. The country was in no way ready to use suddenly such a considerable amount of capital. . . . Therefore, the result was speculation in land, with prices increasing artificially in each transaction; excessive expenditures; and the accumulation of imported goods. . . . When the inevitable hour of payment arrived, crisis ensued. This crisis is now being faced by a decrease in private and public expenditure.[9]

The Oficina de Cambio closed its doors in 1876 and measures were adopted to face the crisis. Public debate around the causes of the crisis and its possible solutions filled many pages of the local papers of the times,[10] but President Avellaneda and his minister Victorino de la Plaza went ahead with their policy of recuperation which slowly reversed the situation for a few years.

The main causes of the monetary question had not yet been solved, however, and monetary chaos persisted throughout the decade.[11] By 1881, several measures had been put into effect in an effort to introduce some rationality into the system. Thus, among the main reforms were the creation of a new unit of gold currency (the golden peso), and the replacement of the old peso moneda corriente by the peso moneda nacional, which was to become the national currency. It was to be issued only by authorized banks under the supervision of the national government, and would be convertible into golden pesos at par. In spite of the importance of these measures, some of their main regulations soon had to be suspended. By 1885, the parity with gold had to be abandoned as a result of a new period of continued deficit in the balance of payments. As Ford explains, at a time when foreign borrowing was declining, the rise of imports and the increase in service charges had a drastic effect on reserves. Therefore, Roca's initial policy of sticking to the gold standard system was replaced by a more flexible approach, but always with the aim of returning to the gold standard in 1887. However, when Juarez Celman came to power in 1886, his administration "then proceeded on its expansionary path, intensifying the foreign-financed investment boom with lax credit and banking policy at home. Exchange rate stability was forgotten."[12]

Thus very briefly the preceding picture has attempted to describe the monetary problems experienced between the 1850s and the 1880s. The explanation of all these problems, however, lies outside the narrow framework of the financial history of the country and in the structural conditions of the Argentine economy and its increasing intervention in and dependency upon the world market.

Chiaramonte has clearly indicated this relationship when referring to the crisis of 1873. He describes what he calls the classic mechanism of the country's balance of payment, resulting from its place in the world market as producer of foodstuffs and raw materials. Periods of expansion, with increased gold reserves due to rising international prices for Argentine staples and to the inflow of foreign capital by way of loans and investment, were followed by periods of crisis and depression, with declining prices and a retracting of investment and loans. In boom years, the chronic deficit of the balance of payment was met with foreign borrowings, but in years of crisis it was aggravated by the withdrawl of foreign credit.[13]

For the pastoral industry, these ups and downs in the monetary field had very specific consequences which the estancieros and exporters were well aware of. From the point of view of landowners, estancieros, and farmers, this alternation between periods of devaluated peso and years of relative stability was lived as one of abundance versus shortage of currency. It has been argued that exporters (both producers of, and merchants in, staples) profited in times of devaluation of the paper peso, as they received their income in gold, but paid their expenses in the national currency.[14] In fact, during most of the pastoral era, periods of devaluated currency prevailed over those of a strong peso (see Table 22), but in critical years of revaluation and scarcity, the voices of the Sociedad Rural rose in discontent. Furthermore, an abundance of currency generally meant easy credit. And if circulation increased greatly throughout the period, so did production and exports, and in times of shortage, money became dear and the cost of borrowing a heavy burden. For an industry like the pastoral one, which depended so much on the availability and cost of capital, the contraction of credit brought hardship and crisis.

In order to understand why the pastoral industry relied so heavily on credit, we must now analyze the capital requirements that the industry faced throughout the period under study, as well as those each pastoralist had to confront when running his enterprise. Finally, we will refer to the financial mechanisms set in motion to cater to those requirements, paying particular attention to the different types of credit available to pastoralists.

Capital Requirements of Pastoralists and the Pastoral Industry

Throughout the period under study in the province of Buenos Aires, the pastoral industry was confronted with different capital requirements, in accordance with the main phases followed by the process of capital accumulation within the industry. The first period of expansion lasted from the late 1840s to the crisis of the 1860s and was characterized by the crossbreeding and multiplication of the flocks, the incorporation of lands previously devoted to other uses or newly opened to productive exploitation, and the adaptation of old estancias to the new activity or the establishment of new stations. The main capital requirements during

Table 22
Value of Pesos Fuertes and Pesos Oro in Paper Pesos—
Annual Average, 1840–1890

Year	Pesos fuertes[a]	Pesos oro[b]
1840	22.00	22.71
1841	18.12	21.43
1842	16.31	16.82
1843	15.60	16.09
1844	13.19	13.60
1845	14.82	15.05
1846	21.30	21.98
1847	20.63	21.25
1848	20.29	21.43
1849	17.50	18.58
1850	14.26	15.00
1851	17.60	18.15
1852	16.13	16.64
1853	18.31	18.88
1854	18.89	19.49
1855	19.97	20.58
1856	20.41	21.07
1857	19.80	20.40
1858	21.39	22.10
1859	20.73	21.31
1860	20.09	20.89
1861	22.70	23.44
1862	23.98	24.83
1863	27.00	25.93
1864	28.81	27.20
1865	27.41	26.23

Sources:

[a]Juan Alvarez, *Temas de historia económica argentina* (Buenos Aires, 1929), 99–100. The table by Alvarez shows the equivalence of peso moneda corriente to peso fuerte, although up to 1863 gold was quoted by the ounce and not by the peso fuerte. Alvarez converted from ounces into pesos fuertes.

[b]The peso oro was established by the law of 1881 as 1.6129 grams of gold of 90% purity. For calculating the value of the golden peso of 1881 in pesos moneda corriente for the period 1840–1880, I have used the figures for the conversion of golden ounces into paper

Table 22 (continued)

Year	Pesos fuertes[a]	Pesos oro[b]
1866	24.35	23.32
1867	24.94	23.32
1868	25.00	24.29
1869	25.00	24.29
1870	25.00	24.29
1871	25.00	24.29
1872	25.00	24.29
1873	25.00	24.29
1874	25.00	24.29
1875	25.00	24.29
1876	28.60	29.14
1877	29.66	29.14
1878	31.95	31.09
1879	32.30	31.09
1880	30.55	27.32
1881	26.93	26.11
1882	25.04	24.35
1883[c]		1.00
1884		1.00
1885		1.37
1886		1.39
1887		1.35
1888		1.48
1889		1.80
1890		2.58

pesos, dividing for each year the amount of paper pesos per ounce by 16.47 (as 1 ounce was equal to 27.064 grams of gold, and therefore 16.47 golden pesos of 1881 were equivalent to 1 ounce). For the table of conversion of ounces to paper pesos, the source used was Olarra Jiménez, *Evolución monetaria argentina* (Buenos Aires, 1971).

[c]After 1882, the paper peso became the peso moneda nacional and the peso oro became the official moneda de cuenta. For 1883 to 1890, quotations are from Alvarez, op. cit., 100.

this period were for acquisition of pure or fine stock to crossbreed with the existing criolla sheep, the installation of basic equipment and buildings, and the purchase of land—mainly public, but also private estates sold to pastoralists. Investment funds at this stage originated mainly outside the industry, and in fact there was a flow of capital from other sectors of the economy toward expanding pastoral activity. This flow was mainly the result of private operations whereby capitalists previously active in other branches (commerce, cattle raising, etc.) invested their earnings in the newly opened trade. However, as we shall see, the mechanisms of credit were already in existence, helping to finance the requirements of the industry.

Internal resources were also used in this expansion. Profit in the pastoral industry derived from three different sources: (1) net profits resulting from the sale of produce (mainly wool, but also skins, tallow, and sheep); (2) natural increase of livestock (which could become part of net profits when sold outside the industry to be consumed as mutton or to be boiled into tallow); and (3) capitalization of land, resulting from the rising price of this resource as a consequence of the increase in land rent. For the industry as a whole in this period net profits resulting from the sale of wool were probably the main internal source of funds used to finance capital requirements. Livestock and land, while fundamental items in the internal process of accumulation, were only secondary sources of investment funds.[15]

After the crisis of the late 1860s, which stopped the process of capital formation and expansion, and probably implied a withdrawal of resources, a second phase of consolidation can be identified between the early 1870s and the beginning of the following decade. This period was characterized by the expansion of sheep raising into new lands bought from the government in the recently opened territories in the southern districts of the province; by the refinement of the existing flocks through widespread and improved crossbreeding, and by the incorporation of wire fencing and other innovations in the physical equipment of the stations, including not only more sophisticated tools but also better housing and buildings.

The dominant capital requirements in this period were for the acquisition of physical assets and for the purchase of lands, particularly those sold by the government in the newly opened areas of the province. Funds for satisfying these needs were provided by sources both inside and outside the industry. Capital from outside the industry was supplied mainly through mortgage financing and other types of credit channeled through different banks and other institutions as well as by more informal networks.

The third phase, which began in the 1880s and continued into the early 1890s, witnessed important transformations in the direction followed by the process of capital accumulation. Within the province of Buenos Aires, the main features of this change were the crossbreeding of the flocks with Lincoln sheep and the gradual displacement of pastoral activities from the counties north of the Salado River, where cattle raising and agriculture started to attract resources previously directed toward sheep breeding, to those south and west of the river.

This process would reach the southern counties as well, but only after the mid-1890s, as up until then flocks continued to expand over that area. For the pastoral industry in the province, crossbreeding probably became the main source of capital requirement, but in the south land and equipment must have had their share in the demand for funds. These were provided from sources both internal and external to the industry, but by the end of the period the outflow of capital must have been more significant than its inflow. In effect, although in certain areas of the country pastoral activities were attracting capital, in the province of Buenos Aires other productive activities started to compete with sheep breeding not only in appealing to outside capital but also in absorbing resources previously devoted to it (land, stations). Therefore, the process of accumulation within the local pastoral industry first came to a standstill and was then followed by a drain of resources which found their way to other sectors of the economy or to other areas of the country.

Although it is possible to outline the main directions followed by the process of accumulation and the principal sources of capital for the different phases found in the period under study, the relationship between the industry's requirements of capital and its own resources remains unknown. Even if the former account of the process suggests that during the first phase of expansion there was an inflow of capital into the industry in the area, and during the third stage a drain of resources toward other sectors, we have found no way of measuring these apparent flows.

Besides these long-term trends in the process of accumulation, the pastoral industry had to face every year the cost of production and commercialization of its produce, which had to be paid generally in advance of the actual sale of the staples abroad and which in the era of pastoral expansion required increasing amounts of cash every season. Therefore, there was a gap in time to be filled between the export season in Buenos Aires and the receipt of the funds from Europe. This gap was mainly filled by credit, and credit was channeled through individual enterprises.

In the case of individual enterprises, it is also possible to distinguish between two types of capital requirements: those posed by the process of investment (setting up, maintaining, and expanding a station) and those which resulted from the cost of operation of estancias and farms. In effect, pastoralists required funds to acquire land, livestock, and equipment, but they also had to obtain the capital necessary to run the station through the year. Wool production was a seasonal activity, and pastoralists had to wait until the shearing was over to realize their main product in the market. But the station had expenses all through the year, and in the high season the cost multiplied while returns were seldom seen until the end of that period. Therefore, besides having to face capital requirements imposed by the process of accumulation at the level of his enterprise, the pastoralist had to solve financial problems which arose every year in relation to the operation of his station.

Both types of requirements were met by drawing on internal resources and by resorting to external sources of funds. Profits accruing from the sale of produce, livestock, and land could be and were invested in the enterprise. Animals could be sold not only outside the industry to be slaughtered and consumed as mutton or boiled into tallow, but also within it as stock to other pastoralists. Although the routine sale of stock was carried out once a year after the shearing season was over, estancieros and farmers resorted to such a practice whenever they were in urgent need of capital and could not acquire it by more convenient means. In the case of land, in certain areas and periods its prices rose so rapidly and so high that on occasions pastoralists chose to sell part of their estates when in need or—less frequently—in order to invest that capital in improving what was left or in acquiring new, larger, and even better-equipped stations in cheaper areas.[16]

Besides drawing on internal resources, pastoralists also resorted to funds originated outside the industry. Investible funds and capital needed to operate the station (circulating capital) were provided in the form of credit by different institutions and persons who gradually defined a complex financial network. In the following pages, the main mechanisms made available by this network to pastoralists in search of credit will be described.

The Provision of Credit

The expansion of the pastoral industry in the 1840s and 1850s was to exert increasing demands on the primitive financial networks that had provided funds to different sectors of the economy during the first decades of the century. Until the early 1850s, banks and other specialized agencies were almost nonexistent, and credit was channeled through a variety of mechanisms.

Import-export houses in Buenos Aires played a key role in this respect, as they had access to credit in Europe, working as they generally did, in close connection with associate or parent houses abroad. Referring to British merchant houses in Argentina, Vera Reber has pointed out: "In the absence of capital reserves, access to credit in the United Kingdom not only provided merchants with the wherewithal for business, but also lubricated the whole structure of Argentine external trade."[17] In effect, mercantile houses in Buenos Aires extended credit through goods on consignment and by discounting letters of exchange and negotiating bills of lading.

Besides these import-export houses, lesser merchants both in Buenos Aires and in the countryside also acted as financial agents, lending money to their customers and making advancements to their suppliers; well-to-do estancieros gave credit to their employees, relatives, and neighbors; and, in general, all those who could afford it entered the speculative business of money-lending, which often left high profits, but could also lead to bankruptcy. Discounting bills became a usual means of advancing capital to third parties, particularly at the height of the export season, and the discount rate was a good indicator of the state of the money market.[18] All these means were used by pastoralists and wool merchants in the early days of the trade, and although specialized agencies and banks came to life during the second half of the century, and particularly after 1870, these old methods persisted throughout the period, coexisting with more formal institutions.

Up to the 1850s the only bank operating in the whole territory of the River Plate Republics was the Banco de la Provincia de Buenos Aires, under its different denominations: Banco de Descuentos, Banco Nacional, Casa de la Moneda, Banco y Casa de la Moneda, and finally—in 1863—Banco de la Provincia de Buenos Aires. In the 1850s, the Confederation government authorized the establishment of the Banco de la Confederación Argentina, which closed only few months after its creation, and the Banco Maua, which only operated from 1858 to 1860.

In 1862, the Banco de Londres y Rio de la Plata opened its doors in Buenos Aires. Although several foreign merchant houses had diversified into banking catering to their merchantile clients, the London and River Plate Bank was the first banking institution based on foreign capital to be established in Buenos Aires. The Commercial Bank of the River Plate followed in 1872, although its shares were primarily in the hands of Anglo–Argentine merchants and only one-third of them were held by bankers abroad. Later on, other such institutions would operate in Buenos Aires, such as the Banco Francés, Banco Alemán Transatlántico, and English Bank of the River Plate.

The year 1872 witnessed the creation of three other important banks in Buenos Aires: the Banco Nacional, a mixed government and private enterprise; the Banco Hipotecario de la Provincia de Buenos Aires, a mortgage bank; and the Banco de Italia y Rio de la Plata, a private institution whose shares were primarily in the hands of Italian residents in Argentina. Furthermore, in the following years, several banks were opened in different provinces of the Interior.

Finally, by 1887, the following banks were operating in the city of Buenos Aires, which was by then capital of the Republic. Created before December 31, 1886, were: Banco Nacional, Banco Provincia de Buenos Aires, Banco de Londres y Rio de la Plata, Banco de Italia y Rio de la Plata, Banco Ingles y Rio de la Plata, Banco

Constructor de la Plata, Banco del Comercio, and Sociedad Mandatos y Préstamos (English). Banks established in 1887 were: Banco Agrícola Comercial, Banco Crédito Real, Banco Buenos Aires, Banco Mercantil, Banco Alemán, Banco Nuevo Italiano, Banco Popular Argentino, Banco Español, Banco Francés.[19]

Institutional development, however, was at first geared to providing pastoralists with short- and at most medium-term credit, and it was only after 1870 that long-term credit on the security of mortgages expanded in Buenos Aires. We shall see how these two types of credit developed throughout the period and how they affected the pastoral industry of the province.

Short- and Medium-Term Credit. One of the main means of short- and medium-term credit in the River Plate was the bill of exchange. Referring to the development of such an instrument in the area, Amaral explains that the bill of exchange only expanded in the River Plate after Independence, when colonial commercial methods and circuits were replaced by new networks of trade.

In Buenos Aires, the bill developed not only as a commercial device but also as an instrument of credit. In effect, a bill was a note addressed by one person (drawer) to another (drawee) ordering him at a certain fixed date in the future to pay a certain sum of money to a specified person (bearer). The bearer presented the document to the drawee who, by accepting it, accepted his obligation to pay the required sum on maturity. Once the bill was accepted, its holder could discount it either at a bank or with a capitalist dedicated to that type of operation. The bank advanced the amount specified in the bill, discounting the interest corresponding to the days left until maturity. At the date fixed by the document, the drawee had to pay the bank who had handled the discount operation the amount stated in the bill. When bills were discounted, they actually became a mechanism of short-term credit, and, as such, they were not always issued as payment orders to the benefit of a third party. Thus, in Buenos Aires, quite frequently bills were drawn for the sole purpose of being discounted, and the document became a device through which one person (who could actually sign as drawer or drawee) could receive short-term credit from the bank, while a second person (who signed inversely) simply endorsed the operation.[20]

The Banco de Descuento first, and subsequently the Banco de la Provincia, institutionalized this system of lending money, which was later put into practice by other banks, such as the Banco Nacional. Actually, the Banco de la Provincia de Buenos Aires was the first commercial bank to be created in Buenos Aires in 1854 with the name Banco y Casa de la Moneda, adopting the former name in 1863.[21] However, since the 1820s several attempts had been made by private capitalists and the provincial government to put into operation a local bank, resulting in the successive creation of the Banco de Descuentos (1822–1826), Banco de las Provincias Unidas (1826–1836), and Casa de la Moneda (1836–1851).

All these institutions were supposed to perform banking functions, as they were established mainly to discount bills and promissory notes, to receive deposits from customers, and to issue paper money. Except the last of them, which in fact after 1838 became only an agency to issue notes and grant loans to the government, the others did actually carry on with their functions—although they ended up with large debts from the government, and had to be dissolved.

When the Casa de la Moneda was established in 1854, its functions were defined in a similar way to those of its predecessors, i.e., discount operations, receipt of deposits, and issuance of notes. However, the bank was soon to prove more flexible than the previous institutions, expanding the scope of its action and attracting customers not only in the city but also throughout the province. In effect, by 1863 the bank had established three branches in the towns of San Nicolás, Mercedes, and Dolores, which were authorized to discount bills (except in gold) and to receive deposits. In 1872 there were already twelve of these branches, and by 1885 there were forty-two. Pastoralists made use of the services offered by the bank, by opening accounts either in the head office, in one of the branches, or both, and particularly by taking advantage of the line of credit offered by the bank through the device of discounting fictitious bills.

The bank could invest its funds in several types of credit operations. As defined by Lamas, these were:

1st. . . . the discount of bills. According to the decree of 27 March

1854 "deposits should be applied exclusively to the discount of bills
. . . with two signatures approved by the Board of Directors, and
for terms not exceeding 90 days." These regulations were modified
by law of 19 September 1860, authorizing the bank to discount bills
with only one signature, and by decree of 25 March 1867, which
authorized the discount of commercial promissory notes on a six
months basis. 2nd. . . . the discount of promissory notes with one
signature on a six months basis with the security of liens on goods
deposited at the customs . . . and the discount of promissory notes
on a one year basis, with only one signature and on the security
of mortgage (law of 1856).[22]

Of these, only the first and the last were practiced by the bank.

The first of the operations described by Lamas—the discount
of bills—became in fact one of the main activities of the bank and
the first type of institutional credit which would be employed by
pastoralists. Lamas goes on to say that by successive renewals,
these bills which were generally to be paid within ninety days
became long-term loans, which he calls *Préstamos de habilitación*.
He adds, "These credit operations did not spring from the law
but rather against the text, or at least against the spirit of the
law. . . . The *préstamos de habilitación* were among the main operations carried out by the bank."[23] According to the law of June
30, 1855, the Board of Directors could fix the maximum credit that
could be given to any particular individual or society. This limit
was fixed first at 300,000$ m/c, then rose successively to 500,000$
m/c, 700,000$ m/c, and 100,000$ fuertes, finally being left to the
discretion of the directors. This ruling was confirmed by the law
of September 1860, which authorized the discounting of bills with
only one signature, thus institutionalizing personal loans, and
accepting these fictitious bills which did not correspond to mercantile operations, as efficient devices to channel credit.

These operations were carried out both by the head office of
the bank in Buenos Aires, and by its branches throughout the
province. The significance of this type of credit operation is hard
to evaluate, but it is possible to follow the expansion of the system
by analyzing the annual movement of drafts, available for the
period 1854 to 1872 in Garrigós,[24] and the annual income received
by the bank for discount operations. Although the information
available refers to real as well as to fictitious bills, a clear pre-

dominance of the second type of operations appears in the credit books of the head office and branches, so that these figures represent mainly the trend followed by those préstamos de habilitación granted by the bank. In the case of discount operations, it is necessary to take into account not only the total amount received by the bank as interest on such operations, but also the discount rate for each year, which varied greatly even from week to week.

Although these figures do not tell us the exact amount of credit granted yearly by the Banco Provincia through the device of these fictitious bills of exchange, they do provide us with an indicator of the trend followed by such operations. And it is quite clear that credit expanded throughout the period, although with certain ups and downs, which generally correspond to periods of boom or crisis in the money market.

These figures say nothing about who benefited from this system, however. In order to find out to what extent pastoralists made use of this type of credit, I have chosen to analyze the discount of fictitious bills in one branch of the Banco Provincia at the heart of the sheep-raising district, from the year of the establishment of the office in 1864 until 1885. Therefore, I have gone through the credit books of the Mercedes branch for the year of its creation—1864—and then for the years 1870, 1875, 1880, and 1885.[25] In these books every drawee of bills appears as heading an account into which all the bills accepted by him as "drawee" are included in one column, while all those in which he acted as "drawer" are listed in another column.

Therefore, most drafts carry two signatures, that of the loan recipient and that of the "drawer," who guaranteed the operation. Generally, the latter was a relative or friend of the former, and sometimes a well-known merchant and/or estanciero of the county. Reciprocity was very frequent, and thus we find a drawee on one page acting as drawer on the next to those who had played the same role when he had discounted his drafts. After the operations with only one signature were authorized by the bank, we see that the larger landowners discounted their drafts with just their name and no second party to guarantee payment.

Besides stating the amount discounted and the date of maturity of the bills, for 1880 and 1885 the books add beside the name of the head of the account his profession and/or the value and characteristics of his property (both real estate and stock). Table 23 sums up the evolution of credit made by the Mercedes branch to pastoralists and landowners in the same county, although the bank also discounted bills presented by persons who lived in neighboring partidos. In Table 24 creditors are divided into five categories, according to the information available for each of them for the year 1880.

These tables suggest an impressive expansion of short-term credit in the province, and point to the importance of such a source of capital for the development of the pastoral industry as a whole. The total amount of these ninety-day drafts discounted by the bank increased sharply between 1864 and 1875, thereafter continuing to expand until 1880, but coming to a standstill in 1885. This credit was obviously open to small pastoralists, who apparently made more use of these drafts than did the larger landowners and estancieros. Owners of under 1,000 hectares of land and flock owners who appear to have no land of their own (tenants, sharecroppers?), represent 80% of the credit holders in 1880 and 76% in 1885, making use of 51% and 56% of the drafts respectively.

This does not mean, however, that larger estancieros did not take advantage of this line of credit offered by the Banco Provincia, but that few of them operated through the local branches, preferring the head office in Buenos Aires—as most of them lived in the city and could benefit from the larger credit channeled through that office. As a test, we selected a few names of well-known pastoralists and looked up their accounts in the credit books of 1870, 1875, and 1880[26] at the head office of the Banco Provincia de Buenos Aires, finding that they not only discounted fictitious drafts with the bank but that the amounts granted to them were larger than those given to smaller pastoralists through the Mercedes branch.

Nevertheless, the importance of such a source of short-term credit for small estancieros, farmers, tenants, and even sharecroppers should not be minimized. Always subject to the onerous conditions of credit imposed upon them by merchants and middlemen, these pastoralists must have benefited from the possibility of discounting bills at the bank, thus obtaining certain sums of capital that might be crucial when buying new stock, at the

Table 23
Drafts Discounted by the Banco de la Provincia de Buenos Aires,
Mercedes Branch, 1864–1885[a]

Year	Number of credit holders	Amount in paper pesos	Equivalence in $oro	Average amount per credit in $oro
1864	17	1,301,000 m/c	47,831	2,814
1870	56	6,457,600 m/c	265,854	4,747
1875	92	20,737,150 m/c	853,732	9,280
1880	267	56,478,038 m/c	2,067,278	7,743
1885	167	2,140,621 m/c	1,562,497	9,356

[a]This table shows the number of drawees who discounted letters in the corresponding year; the amount discounted by each of them has been calculated by adding up the letters accepted by each drawee and their renewals. These letters represent credit extended for 90 days, renewable on maturity, and the sums included in the table result from considering both letters and renewals for each drawee. Includes recipients living in counties of Mercedes and Suipacha, but not those of other counties also served by this branch of the bank.
Source: Original data found in Archivo del Banco de la Provincia de Buenos Aires (ABPBA), Sucursal Mercedes, Libros mayores de crédito, years 1864, 1870, 1875, 1880, 1885.

Table 24
Letters of Credit Discounted by the Banco de la Provincia de Buenos Aires, Mercedes Branch, 1864–1885, by Type of Credit Holder[a] (in $oro)

Year	Large landowners[b]			Medium landowners[c]			Small landowners[d]			Pastoralists with no land[e]			Landowners[f] (no data as to size)		
	Number	Total	Average per credit	Number	Total	Average per credit	Number	Total	Average per credit	Number	Total	Average per credit	Number	Total	Average per credit
1864	2	551	275	—	—	—	4	4,669	1,167	2	1,654	827	9	40,956	4,551
1870	2	14,039	7,109	1	20,708	20,708	16	88,742	5,546	8	32,989	4,124	29	109,376	3,772
1875	3	53,005	17,668	4	224,829	56,207	26	196,488	7,557	25	132,491	5,300	34	246,918	7,262
1880	7	209,846	29,978	18	527,478	29,304	82	579,398	7,066	130	475,739	3,660	30	274,817	9,161
1885	7	222,139	31,734	14	320,907	22,922	63	608,447	9,658	63	267,682	4,249	20	143,320	7,166

[a]Information as to the property owned by drawees was available only for 1880 and 1885 in the credit books of the bank, so this classification is based upon the data found for those years.
[b]Owning above 2,500 hectares.
[c]Owning between 1,000 and 2,499 Ha.
[d]Owning under 1,000 Ha.
[e]Owning sheep but not land. Probably tenants and sharecroppers.
[f]Includes landowners and estancieros for whom no data is available as to the size of their property.
Source: Same as Table 23.

time of the shearing or on account of another such capital-
demanding operation. However, this type of credit did not present a solution to the needs of the pastoralists who wanted to
expand their enterprises, and required a source of long-term and
secure credit.[27] Although these préstamos de habilitación could
be renewed on maturity (ninety days after the operation was
started), and in fact quite frequently were renewed for several
periods, thus becoming medium-term loans, they could also be
claimed by the bank on maturity, leaving the recipient in a very
vulnerable situation. Furthermore, these drafts were subject to
updated interest rates, therefore confronting the holder with several unpredictable situations—as he did not know when he would
have to repay the amount discounted, nor how much he would
have to pay by way of interest upon such a sum. In spite of the
problems that the préstamos de habilitación posed to the less
affluent pastoralists, perhaps Terry was not so far wrong when
he suggested in 1894 that more than half of the estancias of the
province had been developed with funds obtained through the
Banco de la Provincia.[28]

The bill of exchange also was widely used in Buenos Aires in
its original form as a truly mercantile device. As such, drafts were
not only appropriate instruments to facilitate trade but they also
became a means of financing mercantile operations, and indirectly
the pastoral industry as well. The discounting of these bills was
carried out by different banks, merchant houses, and individual
capitalists. In order to understand better how this instrument was
employed by wool merchants and pastoralists, I will analyze the
operations carried out by the London and River Plate Bank in this
respect, between the early 1870s and the late 1880s. The bank was
established in 1862, and it had followed a cautious policy, basing
its operations on an orthodox observation of the safest rules of
banking.[29] As we shall see, the bank was very reluctant to grant
credit on the security of mortgages, and in general was very
careful when making loans or advancements. This attitude was
reflected as well in the policy toward the discount of drafts in
Europe, an activity which by the 1870s represented an important
part of its business in Buenos Aires.

At the height of the pastoral development, most of the documents negotiated through the bank belonged to wool merchants,

who thus obtained the cash necessary to carry out their transactions. The correspondence between the Buenos Aires branch of the bank and head office in London abounds in references to this sort of operation, and although the archives of the institution do not have a specific list of all credits granted in Buenos Aires, by reading through the letters it is possible to gain information on who were the main customers and how drafts were negotiated.[30]

The first significant references to drafts being discounted by wool merchants date from the early 1870s. The main customers were agents or representatives in Buenos Aires of French and Belgian houses, although English and German merchants may be found as well among the drawers and drawees. The head office provided the manager in Buenos Aires with all the necessary information regarding the standing of the customers, while the latter always consulted London before making any significant operation, and often made suggestions as to clients and procedures. The limit for each house or for any particular agent was fixed by the head office, where the conditions of credit were also stipulated.

According to the standing of the drawee, the respectability of the drawer, the amount of business either of them negotiated through the bank, and the interest of the latter in having them as customers, the bank established different conditions for the credit. The limits fixed oscillated between 8,000$oro and 100,000$ oro for any one firm, but most credits were between 20,000$oro and 60,000$oro, generally expressed in francs or sterling. Drafts could be negotiated clean or the bank could ask for the mortgage of shipping documents (bills of lading), and not infrequently advancements were made on account of future drafts. In these operations, the bank's main profit was made through the exchange rate, but sometimes it charged a specified commission, which oscillated between 1/2% and 1%.

The increasing competition from other banks and merchant houses in the negotiation of drafts and passing of exchange plus the expansion of the trade in Buenos Aires led the Bank of London to devise ways of attracting prospective customers. Thus, in September 1879, the head office sent a circular to Continental firms, particularly the ones in French manufacturing districts, informing them that "we are prepared to facilitate the negotiation of their agents' drafts for produce shipments from the River Plate, and inviting them to make the necessary arrangements with us."[31]

The problems that could arise in this type of business are outlined by the manager of the Buenos Aires branch in a letter of November 14, 1879. After referring to the new type of agent sent from Europe to acquire wools, he complained as follows:

> The season commences and one of these men [the wool buyers who come with orders from Europe for the season] commences to execute his orders. Some are authorized to draw to a certain extent in advance of shipping documents, but this is quite exceptional. He commences to buy, but has to pay for his purchases before he can get his Bills of Lading and pass his exchange. He is thus obliged to look for money and he naturally addresses himself to a Bank. He exhibits his so-called letters of credit and the Banker has to decide whether he can advance the money wanted or not. The Banker's task here becomes more difficult than it was when he had to decide as to the credit he could allow a locally established house of standing and reputedly of certain responsibility. He has now to a great extent to trust to the good faith of the man [. . .] If he accords it (credit) he acquires to begin with, a right to the drafts of the party when they are ready and can with reason and fairness, ask a rate of exchange that . . . may be considered advantageous for the banker. But this apparent advantage may not be real. The Banker may not necessitate [sic] to take at the moment, the market may be hardening and it may be his interest to hold off. Still in such cases he must take the bills in order to cover his advances and he must not share too much or complaints will go to Europe, and the Banker, who has really done a great service, will be denounced as an extortionist. This is one of the probable results that might attend such business.[32]

The response of the head office to these complaints instructed the branch manager on how to deal with the new business and reported its decision to eliminate commission charges for future transactions as a means of attracting Continental customers.[33]

Soon enough, French and Belgian firms started to send to the bank their lists of credits issued by them upon Buenos Aires, and the business multiplied. Well-known houses like Kreglinger and Co. of Antwerp, Deherripon of Tourcoing, Charles Masurel of Roubaix, among others, worked with the bank, while important wool buyers like Lix–Klett, Delpech, Soulas, and Boues discounted some of their drafts there. Advances were also made to local wool merchants (some of them estancieros as well) who drew upon European houses (Hale, Duggan, V. Casares, Temperley).

By the mid-1880s the bank found that it could introduce a new line of credit, while limiting the acceptances of drafts. Head Office consulted the Buenos Aires manager on the subject by letter of 28 February 1884:

> I have been designed [sic] to consult you as to the expediency of taking some steps which might reduce the large amounts we are in the habit of holding, of acceptances of the French wool firms. It is certain that these risks are very large, and in comparison to those attaching to our port business in other places, far too heavy.
>
> So many of the drawers on your side are mere agents, without means, that a large portion of the bills we receive consist of, virtually, single named paper; and this class of paper is on the increase, and likely to assume still larger proportions now that the Belgian firms are losing a considerable part of the trade through its diversion to France. The margin of profit derivable from simultaneous drawing and covering operations on your side does not appear to compensate the Bank for these risks, and the Board think it would be highly desirable to see them reduced, if possible.
>
> So however your produce has to be imported by these firms engaged in the trade, there is evidently only one way in which this object could p'haps be attained, and that is by, as it were, compelling them to use Bankers' credits more freely than they do. This course will not be at all to their taste, one reason being the commission it would entail, but it is followed largely in other countries, and we think that this French trade stands alone in the extent of facilities accorded to importers without this class of intervention.[34]

By August, London informed Buenos Aires that the first two letters of credit had been issued. The manager in the branch office was instructed to give preference to the holders of these letters over any others, purchasing their bills against their credits. London added that commission charges to first-class firms should be $3/4\%$ on clean drafts and $1/2\%$ on documentary drafts.[35]

As we can see from this correspondence, the letter of exchange played a very important role in providing short-term credit to the pastoral sector. Besides the London and River Plate Bank, most of the other banks established in Buenos Aires discounted drafts, and so did merchant houses and private capitalists. The draft was a key instrument in filling the time gap between the export season in Buenos Aires and the actual sale of the produce in Europe. Wool merchants and agents in Buenos Aires used the cash obtained through these discount operations to finance their purchases of wool and skins from estancieros, farmers, and middlemen, who in turn used part of this money to cover their own operation expenses.

Besides the draft in its different forms, other mechanisms existed to channel medium- and short-term credit. Overdrafting was one such means, not only provided by banks[36] but also by merchant houses which carried current accounts with their customers and allowed overdrafting.[37] Another possible way of obtaining such credit was by asking advances on the delivery of produce, and in this way most wool buyers, barraqueros, and middlemen became suppliers of capital to the countryside, particularly in the the time of shearing. Local merchants not only advanced cash, but also gave credit on the purchases made by estancieros and farmers, and well-known country merchants like Torroba and Moore gave standing credit to their customers. Finally, capitalists of different sorts (landowners, estancieros, etc.) lent money to pastoralists at fixed rates of interest, not infrequently becoming sources of long-term mortgage credit.

Long-Term Credit. While mechanisms for obtaining short- and even medium-term credit were abundant though expensive, those for acquiring long-term loans were limited to borrowing money on the security of mortgages, and before 1872 even this way generally led to medium-term loans. Mortgaging property in order to acquire capital was a long-standing practice in the River Plate of the 1850s. However, these operations were not carried out in the framework of any banking institution, but rather on an individual basis—money being lent by capitalists at high interest rates (2% to 3% per month) on the security of a mortgage and quite often through contracts of *retroventa*.[39]

When the banking system first started to develop in Buenos Aires, institutions were reluctant to grant long-term loans, although medium-term credit was sometimes advanced on the security of a mortgage. A cautious institution like the London and River Plate Bank was against such credit, although when J. H. Green managed the Buenos Aires branch certain exceptions were made to an otherwise strict policy,[40] clearly stated in a letter of head office to Buenos Aires, dated January 18, 1873:

> Your observations suggestive of the time having arrived when, in view of increasing competition . . . it may be advisable in self-preservation to take mortgages as collateral security, in cases where property is known to exist, and transactions are constant, have met the consideration of the Board, but past experiences of the very tedious nature of all such advances, arising generally from the fact of a mortgage once given being looked upon as an advance, by the taker, for all time so long as interest is paid, would lead the Directors most reluctantly to give their consent for the re-adoption of such course.[41]

Even as late as 1880, mortgages were the basis of medium-term loans. With the creation of the River Plate Trust, Loan and Agency Co. Ltd. in 1881, such practice was institutionalized. Loans in gold currency up to 50% of the value of the property were granted, to be returned in four years at 9% per annum, without commission or amortization scheme. According to Jones, the early loans were used mainly by estancieros of Buenos Aires and Santa Fe to fence and stock part of their estates.[42]

Mortgages on medium-term credit had been handled by the Banco de la Provincia de Buenos Aires since 1856, when authorized to discount promissory notes on the security of mortgages. As Jones points out: "Technically these were short-term loans . . . and a borrower could well find—when he came to renew his loan—that he was required to repay in part or in full, or to pay a higher rate of interest over the next six months."[43]

This type of credit experienced an impressive expansion after 1860, reaching a peak in 1864–1865, and declining at the end of the decade.[44] According to contemporary observers, the institution was not prepared to face the increasing demand for this type of credit, which immobilized too much capital of the bank's portfolio, and in 1871 the provincial government decided to authorize the creation of a new bank to handle mortgage credit, the Banco Hipotecario de la Provincia de Buenos Aires.

The business of the bank consisted in making advances upon the security of mortgages on real estate valued over and above 2,000 pesos fuertes, granting loans to the extent of half the assessed value of the property, and issuing to the mortgagors *cédulas* or bonds at par in lieu of cash. These cédulas, which were issued by series at different times, quoted in the stock market, bore interest rates which ranged from 6% to 8% and were redeemable at 1% to 2% per year. In this way, the bank became a debtor of those who possessed the cédulas, and creditor of those who had mortgaged their property in exchange for the same cédulas. The latter had to pay the bank per year 6% to 8% interest on the loan, 2% amortization, and 1% commission, but they could make these payments in paper currency or in cédulas.

This system became the first mechanism to obtain real long-term credit, and shortly after its creation it became a crucial source of capital for property owners and estancieros. As the Annual Report of the Bank stated in 1876:

> Conditions offered by the Hipotecario . . . are the cheapest and most convenient the market can offer. . . . Who would want to go back to the times when mortgage credit bore an interest rate of 2% per month, with total reimbursement of the loan; to the times of retroventa with immediate expropriation agreed beforehand?[45]

Tables 25 and 26 show the distribution of mortgage credit through the bank from its creation up to 1885. During the first eight years of its existence, under the presidency of Francisco Balbin, the bank followed a cautious policy of lending, contracting credit soon after its creation, on account of the crisis of the early 1870s, and culminating in the suspension of all credit from November 1875 to November 1876. It was also after 1880 that the bank started a clear policy of expansion, increasing the number of loans granted and lowering the interest rate to 6% per annum. Quesada comments upon this policy, pointing out that by 1881 the amounts of the loans and the valuation of property were increased, augmenting the gap between real and assessed value of property.[46]

It was then that bonds became an instrument of speculation par excellence. Ferns has explained how this mechanism worked:

> Once a large number of cédulas were in circulation the advantages

Table 25
Banco Hipotecario de la Provincia de Buenos Aires—
Loans Granted, 1872–1885 (expressed in pesos oro)

	Province		City of Buenos Aires		Counties	
Year	Total number	Total amount	Number	Amount	Number	Amount
1872[a]	813	6,694,330	500	4,355,818	313	2,338,512
1873	608	6,576,447	359	4,110,936	249	2,465,511
1874	317	3,124,247	176	1,985,428	141	1,138,819
1875	336	5,534,087	198	3,378,297	138	2,155,790
1876[b]		279,107		38,808		194,209
1877	190	1,584,449	96	644,329	94	940,120
1878	205	1,502,496	114	532,591	91	969,905
1879	122	919,383	70	328,440	52	590,943
1880	144	966,512	102	469,168	42	497,344
1881	482	3,903,014	307	1,513,432	175	2,389,582
1882	524	4,954,335	281	2,241,410	243	2,712,925
1883	836	12,614,125	347	3,548,503	429	9,065,622
1884	809	13,648,388	367	4,483,206	442	9,165,182
1885	658	10,282,327	346	3,408,256	312	6,874,071

[a]10¹/₂ months.
[b]Credit was suspended between November 26, 1875, and November 26, 1876.
Source: *Memorias del Banco Hipotecario de la Provincia de Buenos Aires,* years 1872–1885.

Table 26
Rural Establishments Mortgaged to the Banco Hipotecario—
1872–1880, 1881, 1882, 1883, 1884, 1885

	Total province				Counties north of Salado River (30 counties)			
Year	Number estancias	Area in Ha	Number chacras	Area in Ha	Number estancias	Area in Ha	Number chacras	Area in Ha
1872–1880	168	927,525	77		52	156,600		
1881	92	635,775	12	3,363	26	74,665	4	1,370
1882	111	528,965	36	8,972	35	87,195	18	5,712
1883	276	1,999,986	41	5,097	76	214,790	20	3,021
1884	184	1,489,229	123	35,360	39	134,816	67	21,361
1885	119	775,954	31	6,164	38	152,472	18	4,012

Source: *Memorias del Banco Hipotecario de la Provincia de Buenos Aires,* years 1881–1885.

of an inflated paper currency to the borrowing classes were very great. And in the Argentine the landed interest, the borrowing interest, was all-powerful politically. Borrowing made it possible for the borrowers to buy more land to offer as security for more debts. Fresh land purchases drove land prices upward and, as land values increased, the capacity of landowners to borrow automatically increased. No more successful instrument of inflationary speculation has probably ever been invented than a cédula. As the time approached for the repayment of loans, borrowers had the most powerful of incentives to drive down the price of cédulas so that they might be purchased as a means of satisfying the Bank.[47]

According to Ferns, money obtained from the sale of bonds was employed mainly in four ways: (1) acquisition of physical assets for rural establishments, (2) conspicuous consumption, (3) speculation in lands, and (4) urban building.[48]

From Table 25 it is clear that during the first four years after the creation of the Banco Hipotecario, the total amount mortgaged in the city of Buenos Aires was higher than that in the countryside, but after 1876 the trend reversed and the countryside (estancias, chacras, and townhouses) attracted the largest amount of credit from the bank. Table 26 shows the number of estancias and chacras that were mortgaged and the area they covered in the province, as well as in the counties north of the Salado River.

For the period under study—the pastoral era—the number of pastoralists who benefited from this type of credit were obviously very few, and these few were generally the better off. The average size of estancias mortgaged was always above 4,700 hectares for the whole province and above 2,400 hectares for the counties north of the Salado (Table 26), although establishments around 1,000 hectares sometimes did receive credit.[49]

Furthermore, by comparing the number of those favored by this type of long-term loans with that of those who had access to short- and medium-term credit through the Banco de la Provincia, we see how limited was the scope of mortgage credit. For example, in the county of Mercedes, up to 1880 only three estancias had been mortgaged to the Banco Hipotecario, while in that year alone 267 landowners and pastoralists had received short-term credit from the Banco de la Provincia. Even for 1885, the figures were 14 and 167 respectively.[50]

It is clear, therefore, that long-term credit granted by the Banco Hipotecario, which was to become an important source of capital for the rural entrepreneurs of the last decade of the century, was only starting to develop in the pastoral era, and that its importance was secondary to other types of credit, mainly short- and medium-term loans channeled through banks and private capitalists.

The Cost of Money

The complex system of financial networks that developed during the second half of the nineteenth century revealed an increasing institutionalization in the form of banks and newly created ways of attracting savings and providing credit, but it did not leave aside the old and more informal mechanisms which persisted and performed an important role throughout the pastoral era. In this sense, commercialization networks played a very important part in providing credit to producers, from the small advances made by acopiadores and pulperos to small wool growers, to the larger amounts lent by merchant houses and brokers to well-known estancieros. But these lenders themselves received credit, through national and foreign banks in the case of the latter, through local capitalists, merchants, and banks in the case of the former. Thus, capital was channeled from different sources toward the pastoral industry, and from it to other sectors of society.

The complex financial networks that were in operation throughout the period offered the pastoral industry a variety of means through which to obtain credit, as well as to channel capital. Yet it proved undoubtedly more efficient in providing short- and medium-term credit than in organizing a stable system of long-term loans. In effect, as we have seen above, only the Banco Hipotecario, created in 1872, was prepared to lend money for long periods and upon the security of mortgage. The rest of the financial mechanisms existing in Buenos Aires were geared to lend for short periods, and generally at periodically updated interest rates.

As a result of this policy, and of the oscillations in the monetary market described earlier in this chapter, the cost of money was extremely variable throughout the period. Figure 8 was built considering annual averages of the discount rate of the Banco de la Provincia for the years 1854–1885, and it is quite clear that during

Figure 8

Discount Rate at the Banco de la Provincia
de Buenos Aires, 1854–1885

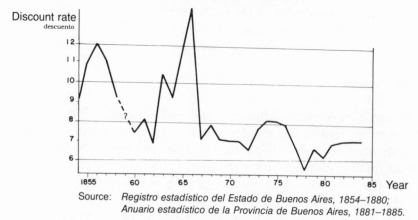

Source: *Registro estadístico del Estado de Buenos Aires, 1854–1880;*
Anuario estadístico de la Provincia de Buenos Aires, 1881–1885.

critical years of credit contraction, the cost of money swelled. Furthermore, the graph shows that the sudden and sharp oscillations of the late 1850s and the 1860s, were followed by a more stable trend, which made the rate of discount oscillate between 6% and 8% in the years 1867–1885. Yet this hides fluctuations occurring within each year, season, month, and even week. Variations during the year could reach up to 100%, as was the case in 1865, when the average annual discount rate for bills in specie was 15% in January, while in November it had dropped to 7½ %.[51] Although this was an exceptional year in terms of the high cost of money and of the oscillations experienced by it throughout the year, generally during the export season (December to May) the rate of discount rose above the values observed during the slack season. In the 1870s and 1880s, however, these differences were generally of one or two points.

These rates represent only one band of the spectrum of rates that could be found in Buenos Aires of the pastoral era, where different banks and private capitalists charged different rates of discount according to their own prestige and interests and to the respectability, importance, and reliability of each customer. Furthermore, there were other types of credit other than the discount of bills, and those also bore their own interest rates. Thus, in times of credit contraction like the years 1864–1866, private capitalists charged between 2% and 2¹/₂% per month to discount for very reputable signatures, but in some cases the interest rate was fixed as high as 40% per year.[52]

In years of abundant credit, like the early 1870s, rates dropped. The usual rates for discounts were 7% to 9%, while 9% to 15% was charged on mortgages. But after these years of credit bonanzas, the cost of money went up again, and in 1875 private discount operations bore an interest of around 20% per annum.[53]

Within this motley picture, well-known and wealthy estancieros and wool merchants were generally in a better position than small wool growers, farmers, and country merchants vis-à-vis these financial networks. In effect, the former not only had better opportunities, information, and connections so as to choose their source of credit, but they also were favored by better rates of interest, and when mortgage credit became available, they rapidly made use of such an advantageous device, borrowing large sums for long periods and at very low interest rates. Furthermore, quite a few of them became an active part of the financial networks by lending money to other pastoralists, and sometimes even receiving savings on deposit, at fixed interest rates, becoming in fact private bankers. When institutionalization speeded up, estancieros and wool merchants were among the organizers, directors, and board members of most banks in Buenos Aires.

The policy of these banks, however, was the result of a conjunction of interests and pressures, and while local institutions like the Banco Provincia or the Banco Hipotecario were very much influenced by the export interests, banks which relied heavily on foreign capital like the Banco de Londres y Rio de la Plata had to attend to many and frequently contradicting voices.

Although estancieros and wool merchants were among the main architects and customers of the financial system of Buenos Aires, small pastoralists were not left out of it, and credit was available to farmers, acopiadores, small estancieros, and even tenant farmers and sharecroppers. However, except for the Banco de la Provincia, most of the institutions created in Buenos Aires during this period scarcely had any mechanisms to cater to these secondary pastoralists, who therefore carried out most of their operations with local merchants, wealthier estancieros, and other capitalists. Therefore, credit obtained by them must have been

generally more expensive and less flexible than that which benefited larger pastoralists. More often than not, they had little choice
when selecting their source of capital, as they had to resort to the
wool buyer who would be ready to lend in advance of the shearing, to the local merchant who agreed to give them standing
credit, or to the landowner or estanciero who kept current accounts with his subordinates. In times of crisis, dependency upon
the source of credit caused more than one bankruptcy and the
hardship of many, but in periods of expansion, it gave the small
wool grower the certainty that he would be able to go ahead with
his enterprise.

Briefly, then, the financial structure of the pastoral era developed from a rather informal network of private capitalists who
carried out some of the main banking functions required by the
society of the River Plate in the early 1850s, to a more complex
and institutionalized system which combined banks of foreign
and local capital with the old mechanisms of lending and saving.
With a greater development of short- and medium-term rather
than long-term credit, and favoring large estancieros and merchants above smaller ones, the system nevertheless ensured the
flow of capital toward the pastoral industry. This flow was subject
to sharp and frequent fluctuations, however, and the money market remained far from perfect throughout the period.

Notes

1. A. G. Ford, *The Gold Standard, 1880–1914: Britain and Argentina* (Oxford, 1962), 90.

2. Ford, op. cit., 94.

3. Ricardo Ortiz, *Historia económica de la Argentina* (2 vols., Buenos
Aires, 1964), I, 143ff; J. C. Chiaramonte, *Nacionalismo y liberalismo económicos en Argentina (1860–1880)* (Buenos Aires, 1971), 50–52; Vera Reber,
British Mercantile Houses in Buenos Aires, 1810–1880 (Ph.D. diss., University of Wisconsin, 1972), 55–56.

4. Thomas Hinchliff, *Viaje al Plata en 1861* (Buenos Aires, 1955), 37.

5. Eduardo Olivera, "Nuestra industria rural en 1866" in *Miscelanea* (2
vols., Buenos Aires, 1910), I, 62–64. Currency in circulation diminished
from around 340.5 million paper pesos in 1862 to 298.5 million in 1865,
while trade expanded. Olivera estimated that around 180 million in

paper pesos was needed to carry on the current transactions in commerce and production. See also Chiaramonte, op. cit., 58–59.

6. See Ortiz, op. cit., 143–53; Chiaramonte, op. cit., 59–60.

7. Ortiz, op. cit., 148–50; Chiaramonte, op. cit., 100.

8. Chiaramonte, op. cit., 116–20.

9. Quoted by Ortiz, op. cit., 148–49 (my translation).

10. Chiaramonte, op. cit., 116–20.

11. For a description of the monetary problems in Argentina in the nineteenth century see, among others, Juan Alvarez, *Temas de historia económica argentina* (Buenos Aires, 1929); J. H. Cuccorese, *Historia de la conversión del papel moneda en Buenos Aires* (La Plata, 1959); Ford, op. cit.; Emilio Hansen, *La moneda argentina, estudio histórico* (Buenos Aires, 1916); Norberto Piñero, *La moneda, el crédito y los bancos en la Argentina* (Buenos Aires, 1921); A. S. Quintero Ramos, *A History of Money and Banking in Argentina* (Río Piedras, 1965); James Scobie, "El desarrollo monetario de la República Argentina durante el período 1852–1865," *Revista del Museo Mitre*, No. 7 (1954); John Williams, *Argentine International Trade under Inconvertible Paper Money, 1880–1900* (Cambridge, 1920).

12. Ford, op. cit., 137.

13. Chiaramonte, op. cit., 112–13.

14. See, for example, Ford, op. cit., chap. 6; Chiaramonte, op. cit., chap. 2 and passim; Ortiz, op. cit., chap. 4 and passim; Reber, op. cit., 54–55.

15. Nevertheless, during the critical years of the late 1860s, many pastoralists resorted to the sale of stock and land to cover their expenditures or pay their debts.

16. Herbert Gibson, *The History and Present State of the Sheep Breeding Industry in the Argentine Republic* (Buenos Aires, 1893), 242–90.

17. Reber, op. cit., 71.

18. William MacCann, *Two Thousand Miles' Ride Through the Argentine Provinces* (2 vols., London, 1853), I, 216ff; *L'Economiste Français*, I (1875), 497; Archives des Affaires Etrangères (France), *Correspondance Commerciale*, Buenos Aires (consular), vol. 7, 1865–1867, f. 124 (1 March 1866); Sixto Quesada, *El Banco Hipotecario de la Provincia de Buenos Aires* (Buenos Aires, 1894), 117; Emile Daireaux, *Vida y costumbres en el Plata* (2 vols., Buenos Aires, 1888), II, 288–89; Estanislao Zeballos, *Descripción amena de la República Argentina* (3 vols., Buenos Aires, 1881/1888), III, 236–37; Reber, op. cit., 50–56.

19. *Censo Municipal de Buenos Aires*, 1887, II, 182. For further notice on banks operating in Argentina in the twentieth century, see, among others, Pedro Agote, *Informe del Crédito Publico Nacional* . . . (Buenos Aires, 1887); Norberto Piñero, op. cit.; J. H. Cuccorese, *Historia económica*

financiera argentina, 1862–1930 (Buenos Aires, 1966), and *Historia de la conversion;* David Joslin, *A Century of Banking in Latin America* (London, 1963); Olarra Jimenez, *Evolución monetaria argentina* (Buenos Aires, 1971); *Banco de Italia y Rio de la Plata, 1872–1972* (Buenos Aires, 1972); Charles Jones, *British Financial Institutions in Argentina, 1860–1914* (Ph.D. diss., Cambridge University, 1973).

20. Samuel Amaral, "Comercio y crédito: el Banco de Buenos Aires (1822–1826)," *América: Revista cuatrimestral de Asuntos Históricos,* 2, No. 4 (April, 1977), 11–12; Andrés Lamas, *Estudio histórico y científico del Banco de la Provincia de Buenos Aires* (Buenos Aires, 1886), 50–52.

21. On the Banco de la Provincia de Buenos Aires and its predecessors, see particularly Octavio Garrigós, *El Banco de la Provincia* (Buenos Aires, 1873); Andres Lamas, op. cit.; Nicolás Casarino, *El Banco de la Provincia de Buenos Aires en su primer centenario 1822–1922* (Buenos Aires, 1922); J. H. Cuccorese, *Historia del Banco de la Provincia de Buenos Aires* (Buenos Aires, 1972); and Susana Ratto de Sambuccetti, *Avellaneda y la nación versus la provincia de Buenos Aires* (Buenos Aires, 1975).

22. Lamas, op. cit., 50–51 (my translation).

23. Ibid., 52 (my translation).

24. Garrigós, op. cit., 216–17, 295–304; *Registro estadístico del Estado de Buenos Aires,* years 1864 to 1880 (section on the Banco de la Provincia); *Anuario estadístico de la Provincia de Buenos Aires,* years 1881 to 1885 (section on the Banco de la Provincia).

25. Archivo del Banco de la Provincia de Buenos Aires, *Libros de crédito,* Sucursal Mercedes, years 1864, 1870, 1875, 1880, 1885.

26. Ibid., Head Office, years 1870, 1875, 1880.

27. See, for example, Quesada, op. cit., 117 and passim; Jones, op. cit., 100 and passim.

28. Terry was national finance minister in 1894. The citation was suggested to me by Charles Jones.

29. On the Banco de Londres y Rio de la Plata see Jones, op. cit., and Joslin, op. cit.

30. *Bank of London and South America* (BOLSA), Letter Books, D1-07 to D1-18, Confidential, Head Office to Buenos Aires, 1866–1885 and D35, Buenos Aires Branch to Head Office, 1871–1882.

31. BOLSA, Letter Book, D1-13, June 1878–May 1880, Confidential, Head Office to Buenos Aires, letter of 11-9-1879.

32. BOLSA, Letter Book, D35, 15-5-1878–11-3-1882, Confidential, Buenos Aires to Head Office, letter of 14-11-1879.

33. BOLSA, Letter Book, D1-13, June 1878–May 1880, Confidential, Head Office to Buenos Aires, letter of 18-12-1879.

34. BOLSA, Letter Book, D1-17, 14-6-1883–3-9-1884, Confidential, Head Office to Buenos Aires, letter of 28-2-1884.

35. Ibid., letter of 9-8-1884.

36. See Garrigós, op. cit., 230, on the practice of allowing overdraft followed by the Banco de la Provincia.

37. For example, Richard Newton carried such an account with Rodger, Best and Co. and Juan Best Hnos. (Sucesión Newton, AGN, Sucesiones No. 7217); Guillermo White with Drabble Hnos. (Sucesión White, AGN, Sucesiones No. 8760); Juan Acebal with Murieta of London, Kleinwort, Cohen y Cia. of London, Baring Bros. of London, and Jose Arellano of Rosario (Sucesión Acebal, AGN, Sucesiones No. 3695); Enrique Bell with Thomas Drysdale (Sucesión Bell, AGN, Sucesiones No. 3971); Claudio Stegmann with Jorge Rick y Cia.—among others—(Sucesión Stegmann, AGN, Sucesiones No. 8225).

38. Several examples of the different forms of credit used by pastoralists may provide a better picture of the financial devices that were in use in the pastoral era. In the 1860s, for example, we find farmer Michael Smith who, at his death in 1867, was found to owe his creditors (Simon Lawler, William Lawler, and Michael Duggan) almost as much as he had been able to accumulate in a lifetime. Cornelio Garahan, who owned 2,700 hectares in Mercedes, by 1863 had borrowed 90,000$ from Duggan and 699,850$ from S. E. Unzué (promissory notes signed when buying a sheep station from him) while he owed Torroba, a well-known merchant in Mercedes, 10,000$ (Sucesión Garahan, AGN, Sucesiones No. 5975); Santiago Wallace, shepherd at Mariano Casares' estancia, who kept a current account with his employer (Sucesión Wallace, AGN, Sucesiones No. 8760); Tomas MacGuire, sheep farmer, who had mortgaged his estate to J. P. Esnaola in 1862 at 80,000$ to be returned in a year at a monthly interest rate of 1.25% paid by semesters in advance (repayment was later postponed until 1869, and the interest rate raised to 2% per month).

In the 1880s we find Juan Fair, shepherd and probably sharecropper, who at his death in 1881 had standing credit with several country merchants and pulperos, like Torroba, Gorostiazu, and Millan. At the other end of the social spectrum, Claudio Stegmann, Jr., in 1888 was in debt with the Banco de la Provincia and the Banco Nacional, for drafts discounted with them, and with the Banco Hipotecario de la Provincia and the Banco Hipotecario Nacional, where he had mortgaged his estates in the country and his houses in the city (Suc. Stegmann (h), AGN, Sucesiones No. 8341). Saturnino E. Unzué in 1886 kept current accounts at the Banco Nacional, Banco de Italia, and Banco de Londres y Rio de la Plata, and at the same time acted as private banker by running current accounts for his clients. He owed the Banco Hipotecario de la Provincia 174,034$ for loans granted upon the security of mortgage on several of his estates (Sucesión Unzué, AGN, Sucesiones No. 8590). Also in the 1880s, the accounts of Terence Moore, well-known country merchant, show that he distributed credit (even mortgage credit) among his customers and ran accounts with several of them, while he kept his own deposits at the Banco Carabassa, Banco de Dublin, and Baring Bros. (Sucesión Moore, AGN, Sucesiones No. 7014).

39. Banco Hipotecario de la Provincia de Buenos Aires, *Memoria anual,* 1880, 9.

40. Jones, op. cit., 100.

41. BOLSA, Letter Book D1-08, March 1872–February 1873, Confidential, Head Office to Buenos Aires, letter of 18-1-1873.

42. Jones, op. cit., 106–9.

43. Ibid., 100.

44. Garrigós, op. cit., 222–23.

45. Banco Hipotecario de la Provincia de Buenos Aires, *Memoria anual,* 1876, 9 (my translation).

46. Quesada, op. cit., 24 (my translation).

47. H. S. Ferns, *Britain and Argentina in the Nineteenth Century* (Oxford, 1960), 370–71.

48. Ibid., 423–24.

49. Banco Hipotecario de la Provincia de Buenos Aires, *Memoria anual,* 1881–1885.

50. Banco Hipotecario de la Provincia de Buenos Aires, *Memoria anual,* 1881, 1885.

51. Chiaramonte, op. cit., 49–50.

52. Archives des Affaires Etrangères (France), *Correspondance Commerciale,* Buenos Aires (consular), vol. 7, 1865–1867, f. 124 (1-3-1866); Chiaramonte, op. cit., chap. 2.

53. *L'Economiste Français,* I (1875), 497.

Epilogue: A New Social Order

The history of Argentina in the nineteenth century is one of a complex, uneven, and often contradictory process of development and consolidation of capitalism. It is also the history of the definite incorporation of the country into the international market as a supplier of food and raw materials and a consumer of manufactures, capital and labor. It has been the purpose of this study to analyze the specific features of that experience in the area that led the whole process—the province of Buenos Aires—and in the period which witnessed the first significant steps toward both systematic accumulation and effective participation in the world market.

One of the main points under debate when considering the history of Latin America has been the relationship between the internal development of each country's economy and society and the external forces set in motion since colonial times. It has been argued that the severance of the colonial link which followed the Wars of Independence was only a prelude to an even greater impact of these external forces upon the region, and that the formation of the liberal nation-states at the end of the nineteenth century came with an increasing dependency upon the metropolis. The nature of dependency, the influence of external forces in stimulating or discouraging the development of capitalist relations of production in each country, the degree of political autonomy that could be achieved in the age of imperialism—all these were central issues in a discussion that started quite early in the history of Latin American scholarly and political debate, and has culminated in the dependency controversies of the last decade. By now the echoes of these debates have died down, but the questions posed then are still relevant. However, the theoretical paths followed in order to search for answers seem to have reached

279

their limit. New questions are now added to the old, and new roads are being explored in order to find the answers.

Given this background of former debates, I have tried to go beyond them in order to respond to some of the old questions with a new perspective. Thus, I have chosen to analyze the way in which capitalism took shape in an area that was increasingly participating in the world market and had a key role in the economic and social history of Argentina. Moreover, I have concentrated on certain aspects of this process—those which in my opinion lie at the core of capitalist accumulation during the second half of the nineteenth century.

After the colonial experience of being mainly an entrepôt in the legal and contraband trade between the Viceroyalty and the rest of the world, Buenos Aires saw its commercial opportunities flourish after Independence, and sought in its rural hinterland the staples to export, hides and salted beef. At this stage, the mercantile classes of Buenos Aires turned their capital and their efforts toward production, and started a long process of capitalist organization of the rural areas. From then on, the international market became the main stimulus for the expansion of production in the province of Buenos Aires, but this in turn was achieved only by the local development of an agrarian structure which could respond to such an incentive.

When, by mid-century, wool became an interesting proposition to the propertied classes of the River Plate as European demand swelled and prices rose, sheep raising came to be the leading sector of the economy, and the process of capitalist organization of production was carried forward with unprecedented vigor. Land and labor were the keystones in this process. Throughout this book, I have described the main steps followed in the private appropriation of the basic natural resources and the development of a land market (chapter 2), as well as in the consolidation of a free labor market (chapter 3). Land and labor were combined by capital in units of production devoted to sheep raising and wool growing, whose main objective was the maximization of benefits, which included not only profits but also rent. Land—fertile and abundant—had been obtained at low prices by those who had privileged access to it through the initial sales and donations made by the state. Availability of this resource contributed to encourage the definition of a pattern of accumulation and of an enterprise organization based upon the extensive use of land and the combination of different forms of labor. Thus, the property of land became an integral part of the business, while rent and profit together ensured the reproduction and expansion of the enterprise in such a way that estancieros could be best described as capitalist landlords (chapter 4).

For decades, estancias were organized following this pattern, although by the 1860s prices of land had started to climb, and investment in land came to represent an increasing share of the total capital necessary to set up a sheep estancia. When prices of wool showed signs of stagnation, this valorization of land meant a decrease in profitability which soon induced capital to try new ways. By the end of the century, beef and grain had replaced wool as the main exports of the country.

The setting up of a complex system of commercial networks was a necessary complement to the productive structures, as the staple had to be marketed in order to realize profits. But the particular form that these networks adopted resulted more from the inherited local trade system and from the international organization of commerce than from any direct influence of the productive system upon it. If anything, it was the commercial circles that generally exerted pressure upon the productive sectors, not only because they were the key to the world market but also because they provided many of the financial services which producers required. At a time when the money market was far from perfect, and when institutional banking was only in its initial stages, commercial networks played an important role as providers of credit and exchange (chapters 6 and 7).

While the producer of wool was at one end of the circuit, the European buyer was at the other, and the existence of the European market was the necessary condition for the expansion of sheep raising in Buenos Aires province. As we have seen, in the long run demand was crucial in the determination of prices (chapter 6), and these in turn regulated the behavior of producers in the River Plate. In effect, prices greatly influenced profitability of rural enterprises, as costs—except financial ones—were a relatively constant item in the calculation of estancieros (chapters 4 and 7). Therefore, when prices of wool remained stagnant for several years and showed no signs of recovery, while demand for other agricultural products such as grain and beef grew in the European markets, and local conditions made its production a feasible enterprise, investments were reoriented toward the new staples, and wool was relegated to the less productive lands.

The European market was not only important to the process of accumulation as a requisite for the expansion of production but also as a source of surplus. In effect, if labor employed in the increasingly capitalist pastoral sector in Buenos Aires was the immediate source of surplus, differential rent probably was not a less important one.

I have not attempted to explore this issue, but have accepted the hypothesis advanced by Laclau and Flichman that the fertility of the soil in the pampean region enabled Argentine staples to be produced at a cost lower than the international market prices, thus giving way to a transference of surplus value in the form of differential rent from buyer to seller.[1] What was the extent of this transference, where did this surplus end up, and how was it distributed are still unanswered questions. It has been argued that in the existence of that international rent lies the explanation of the extraordinary growth of Argentina in the second half of the nineteenth century and the first decades of the twentieth, as well as its subsequent stagnation. Probably, the importance of differential rent in our period did not reach the proportions those authors suggest it would after the 1880s, but nevertheless, as pointed out in the introduction, several facts seem to corroborate its existence, at least in periods of expanding demand and rising prices.

Rent and surplus were the basis of the process of accumulation in the River Plate during the second half of the nineteenth century. Proliferation and improvement of flocks, incorporation of new lands into productive use, multiplication of estancias and farms, population growth and urbanization, and the expansion of transport networks—particularly railways—were only the salient expressions of that process. War, political conflicts, economic crises, and institutional setbacks too often obscured and sometimes seemed to delay accumulation, which was a result both of private investment of profits made by individual capitalists and of increasing public expenditure of social capital.

Throughout this work, it is possible to follow the main aspects of the process of accumulation in the pastoral sector, which was only one—albeit perhaps the most important one in our period—of the areas wherein that process was taking place. And it was not a regular process. On the contrary, it had its ups and downs, knew of crises and booms, and became regressive when other activities developed into more profitable alternatives. The pattern of investment clearly shows the direction followed by capitalism in the River Plate. Although many technological advances were made and flocks were greatly improved, the incorporation of new lands into production was the main object of private and public investment. If production was to remain extensive, land was the basis upon which expansion could take place.

Yet if accumulation was the most obvious destination of surplus, it was not the only one. Although conspicuous consumption would become a striking feature of the gilded age that began in the 1880s, it had found its initial impetus in the previous decades, when an increasing taste for imported consumer goods had begun to change the habits and ways of increasingly larger sections of the population. Outside the capital, this change could be seen more among the rural rich and the professional and commercial circles of the expanding provincial towns, than among the austere small estancieros, farmers and tenants—not to mention the laborers, who did not enjoy such options until much later.

By analyzing the patterns of accumulation and consumption, that of the distribution of the surplus generated by the pastoral industry and the wool trade comes to surface, pressing for an interpretation. Distribution in this period was mainly the result of the dynamics of the private capitalist system, as redistribution through government intervention (taxation, monetary control) was only a secondary way of channeling resources. Actually, after an initial period of appropriation of resources such as land and livestock, the principal mechanism of distribution was the market, but the question of how this was effected has no simple answer. A rapid look at the social transformation of the period may help to detect its main directions.

The relatively simple rural world of the late 1840s, polarized between a group of estancieros and the mass of their workers, still linked by relatively traditional and paternalistic ties, was never so tightly knit as to exclude autonomous and relatively numerous country merchants and shopkeepers, prosperous owners of depots and fleets of carts, independent traders, *troperos,* and *baquianos,* as well as men and women who lived on the margins of rural society. However, forty years later, this assorted variety of characters would seem almost monotonous. The transformation of the estancia organization had introduced new relationships between owner and worker, and while among the former new types appeared and old ones persisted, among the latter, not only their roles were redefined to respond to pastoral requirements but also their old universe was shattered by the introduction of new ways. Thus, the incorporation of the family to the puesto and of increasing numbers of immigrant laborers to the area must have substantially changed the traditional order of the lower classes. Furthermore, a long process of discipline exerted over the labor force and the increasing development of a labor market led to a decline in the paternalistic relations between worker and estanciero, although many traits persisted beyond our period and played an important part in the social organization of the countryside.

Yet the rural society of the 1880s was not only different within the estancia; it was much more so outside its newly fenced boundaries. The vigorous process of expansion had opened up opportunities for native and immigrant farmers and wool merchants; for more pulperos, country merchants, acopiadores, barraqueros, and cart-fleet owners; and for all those who went to live in the growing provincial towns, providing services for the surrounding areas and becoming themselves new sources of demand. Teachers, doctors, shopkeepers, artisans, officers, clerks, and of course workers were settling in increasing numbers in these towns.

The emergence and expansion of all these sectors between the upper ranks of the bourgeoisie and the lower echelons of the working class suggest a new pattern of distribution of wealth. It is not possible, however, to analyze this pattern solely within the rural world, as the city of Buenos Aires was tightly linked to its productive hinterland. As I have mentioned, the decisive steps toward capitalist organization of the rural world had been taken by the mercantile classes of Buenos Aires in the late colonial era. A tight intertwining of interests of port and province was the most immediate result of that initial drive, which found its most ardent enthusiasts among those who embraced both mercantile and agrarian activities, thus becoming merchants *and* estancieros, exporters *and* producers, bankers *and* landowners.

Powerful and rich, this elite of Buenos Aires contributed much to shape the state which in turn was frequently in the hands of governments whose main support came from that elite. Therefore, official policies and private action were frequently directed to the same end, i.e., the organization of a capitalist productive structure. Thus the initial distribution of natural resources, the definition and defense of private property, the imposition of law and order—all tended to contribute, to create, and to protect the capitalist order that was being established in the province. Yet the benefits of this system were not the monopoly of its initiators, and although some of them remained among the powerful and rich throughout the pastoral era, others were to see their fortunes decline, while many newcomers to the propertied classes were on the ascent.

The unprecedented process of accumulation of the pastoral era was vigorous enough to admit the incorporation of new sectors to the capitalist few, but at the same time, it consolidated the fortunes of those few in such a way that by the end of the period most of those who had climbed to the highest ranks of society were there to stay for a long time to come. As we have seen, sheep raising and the wool trade became the launching platform for many of the richest men of the times, a decisive intermediary stage to the owners of many older fortunes, but a fatal step for those who failed to shift to more profitable fields at the right time.

The process of accumulation favored those who had more, but only if they were able to struggle vigorously in the particular capitalist world they were themselves building. Thus, we have seen that the best placed men in the pastoral world were those who were engaged not only in productive but also in mercantile and financial activities, diversifying as well into other fields, and dispersing risks instead of concentrating investment, following a pattern of behavior which was later to prove characteristic of the Argentine ruling class.[2]

In the pastoral age, why was this? Probably because as estancieros, they could obtain the best profits in the prevailing type of enterprises by reducing the cost of money and by increasing income, mainly through better prices, and both could be achieved by strong participation in the urban world of commerce and finance. As merchants and financiers, they could obtain from sheep raising (rightly managed) the highest profits in the market, and the property of land was a secure long-term investment and a source of cash through mortgage credit. On top of these specific reasons, the behavior of the local bourgeoisie was influenced by their own experience: the risky nature of production and trade in a country which based both on primary products could best be countered by a very flexible pattern of investments. Moreover, as the fertility of the soil allowed for high production with low investment and as investment in land could be financed by mortgage, agrarian endeavors did not immobilize large amounts of capital and thus made diversification a feasible and convenient attitude.

Yet, if those few who rose to the top benefited more from the pattern of distribution that was shaping up in this period, other sectors were also receiving part of the surpluses generated or attracted by sheep raising and the wool trade. In this work I have dealt only with those who were rather directly related to both, but most other sectors of society were also touched by this process.

Most sheep estancieros and wool merchants, landowners who let out their lands to pastoralists, many barraqueros and acopia-dores, a number of those involved in the transport of wool, were well placed to receive a good share of the wealth generated within the industry and the trade. Farmers, tenants, and all those self-employed in the handling, transportation, and commerce of wool most likely spent their lives between years of accumulation, when the result of their own work could at least be retained, and years of drain, when their efforts were mostly spent in canceling debts or struggling to ensure the mere survival of their enterprise. Outside this almost purely pastoral world, where the incidence of the wool boom is clear enough, it is harder to detect the direct effect of the trade, although it is evident that the whole of society was involved in the process of expansion unleashed by sheep raising.

Not all this wealth remained within the boundaries of the River Plate society, however, as two main routes were open for resources to flow beyond those limits. On one hand, foreign capital was involved in the pastoral boom, mainly in trade and finance (chapters 6 and 7), and returns from these operations generally found their way to Europe, so that only a part of them was probably plowed back into the local economy. On the other hand, in this period imports increased greatly and foreign investment, particularly in government loans and railways, gained momentum, with both contributing in the long run to the transference of surplus to European countries. However, there was as yet no single country monopolizing this advantage. In effect, if investments and imports came increasingly—though by no means exclusively—from England, direct participation in the wool trade was mainly in the hands of Continental houses—mainly French and Belgian, but also German—though the British were never completely absent from that business and were always quite active in financial endeavors and transport networks.

If it is not hard to observe how the higher classes benefited from economic growth and how a small but far from insignificant sector of society was struggling to participate in the rewards of expansion, it is more difficult to perceive what was happening to the laboring classes.

Within the rural world, native gauchos, herdsmen, and arrieros were giving way to Basque, Irish, and Scotch shepherds and puesteros, who settled with their families in estancias and farms. By the end of the period, Italian agricultural workers also were starting to be a usual sight in the province of Buenos Aires. Day-to-day habits were also changing. At first, as a result of the expansion of sheep raising, mutton rapidly replaced beef in the regular diet of the rural worker. Later on, the introduction of new ways by the immigrants as well as the increasing diffusion of imports through the countryside, helped to bring certain goods and foodstuffs into the homes of the laborers, like *alpargatas*, sugar, tea, or rubber boots.

As mentioned above, the consolidation of a free labor market had reduced the autonomy of the rural poor, who were more and more channeled to wage labor and regular employment. The militarized order of the frontier estancia, always alert to Indian raids, was being slowly transformed into the more pacific toilings of the rural enterprise where women and children took part in work and social life beside the men. Old types became marginal and the blond Irish shepherd with his red-cheeked wife and numerous children employed in an established puesto rose as the antithesis of the solitary and wandering gaucho of the past.

However, this image should not be carried too far or it would become misleading. Actually, the life of large sections of the rural population probably suffered few transformations in this period. In the first place, although immigrants were settling in large numbers in the countryside, they never represented more than 25% of the population of the province. Most workers were natives, and for many of them it is quite likely that the transition between the world of the 1850s and that of the 1880s was at once slow and swift. Probably they had to learn new skills and settle in a puesto or continue to shift from estancia to estancia as shearers or peons; surely they were surprised and later accustomed to new foreign faces and ways; undoubtedly they had to acquire a taste for mutton. But their everyday routine of estancia work was very similar, their entertainments survived almost unchanged, and the quality of their life remained virtually the same.

Not all laborers shared exactly the same world. At least in two ways differentiation among rural workers was increasingly apparent. In the first place, immigrant settlers and native laborers defined two universes of different habits and behaviors, whose edges were blurred, but which found clear expression in the ideological images that each one built of the other. Thus, while Irish shepherds became expert riders, Basque settlers came to like *mate*, and both shared work with the natives. They tended to maintain national solidarities which rose above class barriers, and which gave them a sense of superiority and security that helped them in their integration into the new society.

Secondly, certain stratifications among rural laborers took place in the boom years of the pastoral era. The opening up of opportunities for workers engaged in sheep raising—high wages, aparcería—created a new and wider social spectrum and even expectations of a possible "career," where a laborer could start at the bottom as temporary peon and rise to established shepherd, eventually developing into an independent tenant or farmer. These expectations were more common among immigrants than among native workers, as the living example of few of the former who had actually followed that career with success veiled the reality of many more who remained as wage laborers all their lives. In fact, whatever possibilities arose in the boom years of sheep raising, they were obviously diminishing by 1880.

But by 1880 many other things had changed in Buenos Aires. The boom era of sheep raising had seen growth and risk go hand in hand. The possibilities of becoming wealthy after a few good years in sheep raising were parallel to those of going bankrupt in an even shorter span. Land was still cheap, labor was well paid, and opportunities arose for many newcomers, although not a few of them were defeated by natural forces, market conditions, or expensive credit. The initial process of accumulation was brutal to many, but dynamic enough to be open to more than the few already at the top when the process started.

By 1880, it was perhaps not so common to go bankrupt, but nor was it likely to rise from rags to riches in the pastoral sector. The slackening pace of expansion, the increasing prices of land, a tighter organization of credit and trade—all contributed to make the pastoral sector more secure and stable, but also less dynamic. The bourgeoisie who had conquered the higher ranks of the social order were better protected to pursue systematic accumulation, and the consolidation of the capitalist order was completed.

This study of the process of accumulation during the pastoral era has been necessarily incomplete. On one hand, I have made only side references to certain very important aspects of the process which require further research, such as the nature of the socio-political order and of the state during that period. On the other, I have limited the analysis to the pastoral sector in one area, leaving aside the rest of society, and referring only to how surplus was generated and accumulated within that sector, and how distribution affected those directly involved in it.

The period of systematic accumulation unleashed by sheep raising did not end with the decline of the wool trade or the migration of sheep to far-off areas. On the contrary, an unparalleled process of expansion was witnessed by the province of Buenos Aires following the pastoral era and lasting until the first decades of the twentieth century. Only then did the pattern of accumulation barely sketched in the 1860s but already clearly defined in the 1880s prove unable to ensure systematic growth and expanding wealth. But, by then, a new society had developed in the River Plate and the history of its transformation is far beyond the purposes of this work.

Notes

1. Ernesto Laclau, "Modos de producción, sistemas económicos y población excedente: aproximación histórica a los casos argentino y chileno," *Revista Latinoamericana de Sociología*, 5 (1969), 276; Guillermo Flichman, *La renta del suelo y el desarrollo agrario argentino* (México City, 1977).

2. Jorge F. Sábato, *La clase dominante en la Argentina moderna: Formación y características* (Buenos Aires, 1988).

Appendix

Main Laws on Public Lands in Buenos Aires Province

During the second half of the nineteenth century both the national and the provincial administrations passed a number of laws which became the main legal instruments through which public lands were transferred to private hands. The most important of those measures are the following:

(a)   Law of August 1857 authorizing the sale of 100 sq. leagues (270,000 hectares) of land north of the Salado River at a price of $200,000 per square league (equivalent to 3.63$oro per hectare), the product of which would be applied partly to finance a loan granted to the province by the British, and partly given to the Banco Provincia and to the local governments of those counties where the sales would take place. Under this law, the government issued 349 legal titles, most of them for holdings not over 1 square league, and until 1863, it had sold land for $15.5 million (approximately 760,000$oro).[1]

(b)   Law of October 1857 establishing the leasing of public lands that until then had been held under the enfiteusis system, and giving priority to those who had been occupying the land, and who could prove to have settled therein. Under this law, 534 leases were granted—179 within the frontier, the rest beyond—adding up to 4 million hectares. The plots leased varied in size, but 70% of them were between 2 and 3 square leagues and 19% around 6 square leagues. Jacinto Oddone has revealed the names of most of the tenants favored with leases, showing that most of them were among the largest landowners of the province. However, Valencia de Placente has suggested that although a large number of leases were transferred from the original holders to third parties, thus giving way to the concentration of land, this process was not so acute as Oddone has described.[2] All land rented out under this law was later sold by the government under the laws of 1867 and 1871. According to

291

Bejarano, by 1863 the leases paid had produced no less than 10 million pesos (385,000$oro).[3]

(c)   Law of 1859 allowing for the sale of another 100 square leagues—this time south of the Salado River—and establishing no limit as to the amount each person could purchase. Income produced under this law would be applied mainly to finance the fiscal deficit.[4]

(d)   By the Law of 1864 the government called for the sale of all lands available within the frontier, as it was defined in 1858. This meant that 800 square leagues of land (over 2 million hectares) were to be sold, giving priority to the occupants (subtenants and tenants), but establishing prices that were considered too high at the time: $400,000 per square league (5.44$oro per hectare), north of the Salado, and $200,000 (2.72$oro per hectare) south of the Salado. This law was strongly criticized by estancieros like Olivera, who claimed that in the midst of the mid-1860s crisis the fixing of such high prices for public land would only aggravate the problems the countryside was going through.[5]

(e)   By 1867 Avellaneda was Minister of Government of the province, and he found this an excellent opportunity to further his views regarding the need to consolidate private property of land. He set out to elaborate a law that would promote the transference of public land to private hands, by offering tenants the possibility of purchasing—at the end of the period of the original lease—the land they had been renting at prices that seemed more reasonable than those established by the Law of 1864. Land that was not claimed by its occupants would be sold in public auction. This law proved quite successful from the point of view of the Treasury, but critics of the project found the price-fixing scheme too rigid to satisfy the market, and the system of payment in five annuities too pressing for any relatively small tenant who wished to purchase the land.[6]

The government granted 494 legal titles under this law which, adding up to almost 1 million hectares, was half the extension it had expected to sell. Although 75% of the holdings sold were not over 1 square league in size, 20% were between 2 and 4 square leagues, and 5% between 5 and 6 square leagues. Also, a few families were able to concentrate larger extensions by accumulating plots.[7]

(f)   By 1871, another law established the sale of lands beyond the frontier, establishing no limit as to the amount of land each person could purchase. Under this law the government issued 590 titles adding up to over 3.7 million hectares, which meant an income of almost 3 million golden pesos for the Treasury. These sales resulted in the consolidation of latifundia beyond the frontier, an area which until then had been largely kept in the hands of the state.[8]

(g) Finally, the Law of 1878 put up for sale all public land within and beyond the frontier. The actual occupants of any holding, if interested in purchasing it, had to claim their rights and could buy up to 8,000 hectares of that land. The rest would be auctioned, with facilities to pay in eight years' time. The purchasers were authorized to acquire up to 30,000 hectares in a single section, as defined by the law. There is no doubt that this law was meant to favor the large landowners who, together with speculators, profited from the possibility of concentration.[9]

Notes

1. Miguel Cárcano, *Evolución historica del régimen de la tierra pública, 1810–1916* (Buenos Aires, 1917), 117; M. E. Valencia de Placente, *La política de tierras públicas después de Caseros, 1852–1871* (Ph.D. diss., University of La Plata, 1983).
2. Jacinto Oddone, *La burguesía terrateniente argentina* (Buenos Aires, 1967); Valencia de Placente, op. cit.
3. Manuel Bejarano, "Inmigración y estructuras tradicionales en Buenos Aires (1854–1930)," in T. Di Tella and T. Halperin Donghi (eds.), *Los fragmentos del poder* (Buenos Aires, 1968), 86.
4. Cárcano, op. cit., 121.
5. Olivera, "Nuestra industria rural en 1866," in *Miscelánea* (Buenos Aires, 1910), 70–83.
6. Cárcano, op. cit., 248; *Le Courier de La Plata*, 22 June 1886.
7. Valencia de Placente, op. cit.
8. Cárcano, op. cit., 249; Valencia de Placente, op. cit.
9. E. Barba, et al. "La campaña del Desierto y el problema de la tierra; La Ley de 1878 y su aplicación en la Provincia de Buenos Aires," in *Segundo Congreso de Historia de los Pueblos de la Provincia de Buenos Aires* (La Plata, 1974), 230; Bejarano, op. cit., 98; Cárcano, op. cit., 253.

Glossary

Acopiador de frutos

Middleman who collected produce in the countryside to sell later at profit.

Agarrador

During the shearing, men who were in charge of catching the sheep from the flock to pass them to the shearers.

Agregado

Person authorized to raise a hut free of charge on the boundaries of an estancia. Sometimes, stranger or distant relative adopted into a household.

Agropecuario

Agricultural (relative to cultivation and cattle raising).

Animal de avanzada

Cattle used to prepare the ground before sheep were introduced.

Aparcero

Sharecropper.

Arrendamiento

A land-leasing contract, or the money paid as rent.

Arriero

Muleteer.

Arroba

Weight of twenty-five pounds.

Atador

During the shearing, men who collected the fleeces and made bundles out of them.

Barraca

Warehouse.

Barraquero

Barraca owner.

Bicho

Small grub or insect. In the Argentine countryside, it is used as equivalent to *animal*.

Boleto de premio
 Certificate stating right of property over a certain amount of fiscal land granted by the government to reward those who had rendered service in the wars against the Indians or in other endeavors. Boletos could be presented by the holder to the authorities in order to claim for the corresponding tract of land or, as was frequent in the case of soldiers, sold in the market for cash.

Bota de potro
 Boots made with the skin of a horse leg.

Borrega
 Lamb not yet a year old.

Cabaña
 Establishment devoted to raising pure breeds and to crossbreeding.

Cabecera de partido
 Main city of a county.

Capataz
 Overseer.

Cédulas hipotecarias
 Mortgage securities.

Chacarero
 Farmer.

Chacra
 Farm.

Chiripá
 Sort of trousers used by pampean gauchos.

Comisario
 High ranking police officer; chief of police station.

Comandante
 Commanding officer of armed forces.

Criollos
 When referring to sheep, indigenous breeds.

Desmerinización
 Process through which the predominantly merino flocks in Buenos Aires province were replaced by crossbred Lincoln flocks, in the 1880s.

Enfiteusis
 Leasing of public lands under a special law that allows for long-term contracts at low prices.

Enfiteuta
 Person holding land in enfiteusis.

Esquila
    Shearing.
Fería
    Open market.
Galponero
    Worker in charge of the sheds.
Ganadero
    Stock-farmer, person devoted to cattle raising, grazier.
Grasería
    Establishment where tallow is produced.
Hacendado
    Hacienda owner.
Hacienda
    Large estate. In the province of Buenos Aires, *cattle*.
Huachillero
    Shepherd of the Peruvian highlands.
Jornalero
    Day laborer, unskilled.
Juez de Paz
    Justice of the Peace.
Latifundista
    Owner of large tracts of land.
Letra de Tesorería
    Draft issued by the Treasury.
Lienzo
    Board used for fencing.
Mediería
    Sharecropping contract in which the sharecropper receives half
    of the produce.
Merino
    Breed of sheep with very fine wool.
Ovejas al corte
    Sheep of all sorts found in a flock, previous to any selection.
Papeleta de conchabo
    Job certificate.
Partido
    Administrative units or counties of Buenos Aires province.
Peon
    Unskilled wage laborer. In Argentina, the word does not imply
    servitude or bondage.
Puestero
    Man in charge of a puesto.

Puesto

In an estancia, area under charge of a man who takes care of a flock of sheep (or of a herd of cattle), including his living place, usually a hut (also known as puesto), corrals, and fences.

Pulpería

General store and tavern in the countryside.

Quinta

Periurban small establishments devoted mainly to growing vegetables. In an estancia, area devoted to growing vegetables for internal consumption.

Rancho

Hut made of adobe.

Saladero

Factory for salting beef.

Tercería, tercianería

Sharecropping contract in which the sharecropper receives a third of the produce.

Tropero

Muleteer.

Vagancia

Vagrancy.

Vagos y malentretenidos

Vagrants, loafers. The expression was used to refer to persons who had no established job or source of income.

Vale

Voucher.

Vizcacha

A South American rodent.

Yerba

Used to prepare *mate*, an infusion drunk in countries of the Southern Cone of Latin America.

Bibliography

Guides and Bibliographies

Bagú, Sergio. *Argentina, 1875–1975: Población, economía, sociedad, estudio temático y bibliográfico*. Mexico, 1978.

Coppejans-Desmedt, H. *Guide des Archives d'Entreprises conservées dans les depôts publics de la Bélgique*. Bruxelles, 1975.

Halperin, Tulio. "Argentina." In R. Cortés Conde and S. Stein (eds.), *Latin America: A Guide to Economic History, 1830–1930*. Berkeley, 1977.

Handbook of Latin American Studies. Cambridge, 1936–86.

La historia económica en América Latina. 2 vols., Mexico City, 1972.

Liagre-Baerten. *Guide des sources de l'histoire d'Amérique Latine conservées au Belgique*. Bruxelles, 1967.

Trifilo, Samuel. *La Argentina vista por viajeros ingleses, 1810–1860*. Buenos Aires, 1959.

Wilgus, Alvah. *Latin America in the Nineteenth Century: A Selected Bibliography of Books on Travel and Description Published in English*. Metuchen, N.J., 1973.

Primary Sources

Archives

Archivo General de la Nación (AGN), Argentina:

Sucesiones: Juan Acebal, 1878 (No. 3695); Enrique Bell, 1862 (3971); Jose Mariano Biaus, 1870 (4026); Jose Dowling, 1866 (5426); Juan Fair, 1881 (5795); Cornelio Garahan, 1863 (5975); Federico Gándara, 1882 (6119); Elías Girado, 1867 (5995); Federico Girado, 1882 (6119); Felipe Girado, 1863 (5976); Francisco Girado, 1867 (5993); Laureana Guevara, 1865 (5981); Patricio Lynch–Isabel Zavaleta, 1881 (6624); Tomas MacGuire, 1867 (6852); Miguel Murphy, 1865 (6842); Guillermo Mooney, 1867 (6852); Terencio Moore, 1885 (7014); Ricardo Newton, 1868 (7217); Peter Sheridan, 1861 (8184); Tomas Sillitoe, 1887 (8333); Michael Smith, 1869 (8225); Claudio Stegmann, 1869 (8225); Claudio F. Stegmann, 1888 (8341); Bartolomé Sueldo y Gómez, 1858 (8184); Saturnino Unzué, 1854 (8578); Saturnino E. Unzué, 1886 (8590); Samuel Wheeler, 1865 (8760); Guillermo White, 1866 (8760); Agustín Zemborain, 1886 (8809).

Policía, Tabladas, Corrales y Mercados. Sala 10, 3-2-6 (1856), 32-5-3 (1861–63).

Archivo Senillosa. Sala 7, 2-6-13, 2-5-9, 2-5-11 (1842–1865 approx.).

Libros de Contribución Directa, Campaña. Sala III, 33-4-18 (1845), 33-5-5 (1850), 33-5-14 (1855), 33-8-21/22 (1860), 32-11-3/3/4/5 (1865), 33-8-28/32 (1867).

Protocolos Notariales. 1850 to 1880.

Libros de cuentas corrientes de Pastor Obligado, from 1848. Colección Biblioteca Nacional 800–801.

Tribunales de comercio de Buenos Aires. Lawsuits cited throughout the text, 1850–1870.

Archivo del Banco de la Provincia de Buenos Aires:

Sucursal Mercedes, *Libros de crédito*. Years 1864, 1870, 1875, 1880 and 1885; Books No. 2943–2955.

Casa Central, *Libros de crédito*. Years 1870, 1875, 1880; Books No. 799–807.

Archives des Affaires Etrangères, France:

Correspondance commerciale. Buenos Aires. Vols. 3–20, 1838–1900.
Correspondance commerciale. Anvers. Vols. 14–15, 1862–1871.

Archives Nationales, France:

Douanes. Produits et dépouilles d'animaux: 1848–1910 (F^{12} 6142 and 6843).
Relations commerciales avec les pays etrangères. Confédération Argentine, 1857–1899 (F^{12} 6468).
Buenos-Ayres, 1868–1885 (F^{12} 7041) and 1886–1904 (F^{12} 7042).

Bank of London and South America (BOLSA), documents deposited in University College, London:

Letter Books. Confidential, Head Office to Buenos Aires, 1871–1885 (Books D1-07/18).
Letter Books. Confidential, Buenos Aires to Head Office, 1871–1884 (Books D35-1/5).

Private Archives of the Etablissements G. Cormouls-Houlès, Père et fils, Mazamet, France. Correspondence:

Livres Particuliers. Buenos Ayres. Vols. for 1872–1878, 1879–1884, 1884–1887, 1887–1889, 1889–1892.

Lettres de M. Gaston Cormouls-Houlès. 1873–1889.

Lettres de M. Gaston Cormouls-Houlès. "Buenos Ayres–Particulier," 1889–1892.

Private Archives of Ronald and Rodger, Liverpool, G.B.:

Circulars published by the firm Hughes and Ronald on the state of the wool trade in the world between January 1884 and December 1859.

Public Records Office, London:

Foreign Office, F.O.6/174–408, 1853–1889.

Published Sources and Contemporary Works:

Government Publications:

Argentina, República. *Estadísticas de las Aduanas de la República Argentina, 1870.* Buenos Aires, 1871.

———. *Estadística general del comercio exterior de la República Argentina.* Vols. of 1871, 1872, 1873, 1874. Buenos Aires, 1872, 1873, 1874, 1875.

———. *Estadística de la República Argentina. Cuadro general del comercio exterior.* Vols. of 1875, 1876, 1877, 1878, and 1879. Buenos Aires, 1876, 1877, 1878, 1879, 1880.

———. *Estadística del comercio exterior y de la navegación interior y exterior de la República Argentina.* Vols. of 1880 and 1881. Buenos Aires, 1881, 1882.

———. *Estadística del comercio y de la navegación de la República Argentina.* Vols. of 1882 to 1890. Buenos Aires, 1883–1891.

———. *Primer censo de la República Argentina, 1869.* Buenos Aires, 1872.

———. *Registro Estadístico de la República Argentina.* Vols. 1864–1868. Buenos Aires, 1865–1869.

———. *Registro de la República Argentina, que comprende los documentos expedidos desde 1810 hasta 1890.* 13 vols. Buenos Aires, 1881–1899.

———. *Resúmenes estadísticos retrospectivos.* Buenos Aires, 1914.

———. *Segundo censo de la República Argentina, 1895.* Buenos Aires, 1898.

Buenos Aires, Provincia de. *Antecedentes y fundamentos del Código rural.* Buenos Aires, 1864.

———. *Anuario Estadístico de la Provincia de Buenos Aires.* Vols. of 1881–1890. Buenos Aires, 1882–1891.

————. *Censo agrícolo-pecuario de la Provincia de Buenos Aires, 1888.* Buenos Aires, 1889.

————. *Censo general de la Provincia de Buenos Aires. Demográfico, agrícola, industrial, comercial, etc., verificado el 9 octubre 1881.* Buenos Aires, 1883.

————. *Código rural de la Provincia de Buenos Aires.* Buenos Aires, 1865.

————. *Código rural de la Provincia de Buenos Aires,* ampliado con modificaciones introducidas al mismo por la Sociedad Rural Argentina. Buenos Aires, 1870.

————. *Registro estadístico del Estado de Buenos Aires.* Vols. of 1854–1880. Buenos Aires, 1855–1881.

————. *Registro gráfico de propiedades rurales de la Provincia de Buenos Aires.* Buenos Aires, 1836.

————. *Registro gráfico de las propiedades rurales de la Provincia de Buenos Aires.* Buenos Aires, 1864.

————. *Registro gráfico de las propiedades rurales de la Provincia de Buenos Aires.* Buenos Aires, 1890.

————. *Reglamento de las municipalidades de campaña, 1856.* Buenos Aires, 1861.

————. *Tierras públicas. Recopilación de leyes, decretos, etc. 1810–1895.* La Plata, 1896.

Buenos Aires, Provincia de, Banco Hipotecario. *Memorias anuales.* Vols. of 1872–1885. Buenos Aires, 1873–1886.

Buenos Aires, Provincia de, Banco de la Provincia. *Leyes y decretos desde 1854 a 1887.* 2 vols. Buenos Aires, 1888.

Buenos Aires, Provincia de, Dirección de Ferrocarriles. *Memorias.* Vols. of 1885–1888.

Buenos Aires, Provincia de, Ferrocarril del Oeste. *Memoria del Directorio.* Vols. of 1873, 1876, 1877, 1878, 1881, 1882, 1883. Buenos Aires.

France, Administration des Douanes. *Tableau général du commerce de la France avec ses colonies et les puissances étrangères pendant l'année. . . .* Vols. of 1840–1890. Paris, 1841–1891.

France, Ministère de l'Agriculture et du Commerce. *Documents sur le commerce extérieur.* Paris, 1843.

Great Britain. *Parliamentary Papers.* Vols. of 1840–1900, part. Commercial Reports and Accounts and Papers.

Royaume de Belgique. *Recueil consulaire.* Vols. I–XXXV. Bruxelles, 1856–1880.

————. *Tableau général du commerce de la Bélgique avec les pays étrangères pendant l'année. . . .* Vols. of 1840–1890. Bruxelles, 1841–1891.

Contemporary Works:

Araoz, Toribio. *La campaña en la actualidad bajo sus relaciones con la agricultura, industria y comercio.* Buenos Aires, 1859.

Armaignac, Henry. *Voyages dans les pampas de la Rep. Argentine*. 2 vols. Tours, 1883.

Arnold, Samuel. *Viaje por América del Sur, 1842–1848*. Buenos Aires, 1951.

Aurignac, R. d'. *Trois ans chez les Argentines*. Paris, 1890.

Avellaneda, Nicolás. "Estudios sobre las leyes de tierras públicas." In *Escritos y discursos*. 6 vols. Buenos Aires, 1910. Vol. 5.

Beck Bernard, Charles. *La République Argentine*. Lausanne, 1865.

Blockhuys, J. E. "Le marché de laines à Anvers." In *Rapports commerciaux*. Vol. I, pp. 181–193. Antwerp, 1881.

Bonwick, J. *Romance of the Wool Trade*. London, 1887.

Brougnes, Augusto. *Cuestiones financieras y económicas de la República Argentina*. Buenos Aires, 1863.

————. *Extinction du paupérisme agricole par la colonisation dans les provinces de La Plata—Amérique du Sud—suivi d'un aperçu géographique et industriel de ces provinces, avec deux cartes*. Bagnères-de-Bigorre, 1854.

Bruyssel, E. *La République Argentine. Ses ressources naturelles*. Bruxelles, 1888.

Caudelier, C. *La vérité sur l'émigration des travailleurs et des capitaux belges*. Bruxelles, 1889.

Chaubet, C. "Buenos Ayres et les provinces argentines," *Revue Contemporaine*, 29 (1856–1857), 233–261 and 473–500.

Coni, E. R. *Die Provinz Buenos Aires*. Zurich, 1884.

Coninck, Pierre. *Le Havre*. Le Havre, 1869.

Cormani, G. *Argentina: Guida per l'emigrazione*. Milan, 1888.

Crawford, R. *Across the Pampas*. London, 1884.

Daireaux, Emile. *Buenos Aires, La Pampa et la Patagonie*. Paris, 1877.

————. *Vida y costumbres en el Plata*. 2 vols. Buenos Aires, 1888.

————. "La colonie française de Buenos Aires," *Revue de deux mondes*, 3rd. series, 65 (1884), 879–907.

Daireaux, Godofredo. *Cada mate, un cuento*. Buenos Aires, 1945.

————. *Costumbres criollas*. Buenos Aires, 1915.

————. *La cría de ganado en la estancia moderna*. Buenos Aires, 1904.

————. *El hombre dijo a la oveja. Fábulas argentinas*. Buenos Aires, 1905.

————. *Tipos y paisajes criollos*. Buenos Aires, 1903.

Darbyshire, Charles. *My Life in the Argentine Republic*. London, 1918.

Delpech, Emilio. *Una vida en la gran Argentina*. Buenos Aires, 1944.

Dillon, L. A. *Twelve Months Tour in Brazil and the River Plate with Notes on Sheep Farming*. Manchester, 1867.

De Harven, E. *Le marché de laines à Anvers*. Antwerp, 1879.

Dodds, James. *Records of the Scottish Settlers in the River Plate and their Churches*. Buenos Aires, 1897.

Faure, Felix. *Le Havre en 1878*. Le Havre, 1879.

Garrigós, Octavio. *El Banco de la Provincia*. Buenos Aires, 1878.

Gibson, Herbert. "La evolución ganadera." In *Censo Agropecuario Nacional: La ganadería y la agricultura en 1908*. Monografías. Vol. III. Buenos Aires, 1909.

———. *The History and Present State of the Sheep Breeding Industry in the Argentine Republic*. Buenos Aires, 1893.

Gilmour, Graham. *The Argentine Republic (South America) as a Field for British Emigration*. Glasgow, 1865.

Gobineau, Comte de. "L'émigration européen dans les deux Amériques," *Le Correspondant*, 89 (1872), 208–242.

Guilaine, L. *La République Argentine*. Paris, 1889.

Hadfield, William. *Brazil and the River Plate in 1868 Showing the Progress of those Countries since his Former Visit in 1853*. London, 1869.

———. *Brazil and the River Plate, 1870–1876*. London, 1877.

Hannah, J. *Sheep Husbandry in Buenos Aires*. Buenos Aires, 1868.

Harratt, Juan. *Estudios prácticos sobre la cría y refinamiento del ganado lanar*. Buenos Aires, 1885.

Henkel, W. "Die Wolleproduktion der La Plata–Staaten," *Die Welthandel*, 2 (1870), 573–575.

Hernandez, José. *Instrucción del estanciero*. 2nd ed. Buenos Aires, 1964.

Hinchliff, Thomas. *Viaje al Plata en 1861*. Buenos Aires, 1955.

Hutchinson, Thomas. *Buenos Ayres and Argentine Gleanings: With Extracts from a Diary of Salado Exploration in 1862 and 1863*. London, 1865. Spanish ed. *Buenos Aires y otras provincias argentinas (1860)*. Buenos Aires, 1945.

Jefferson, Mark. *Peopling of the Argentine Pampa*. Port Washington, 1971.

Jerdein, Arthur. *The Argentine Republic as a Field for the Agriculturalist, the Stock-farmer and the Capitalist*. London, 1870.

Jeudy, R. "Voyage à la République Argentine (1876–77)," *Revue de Géographie*, 2 (1877), 264–273, 327–338, 414–423.

King, Anthony. *Tales and Adventures in the Argentine Republic, with the History of the Country and an Account of its Conditions before and during the Administration of Governor Rosas* (London, 1852).

———. *Twenty-four Years in the Argentine Republic*. London, 1846.

Konig, A. *A través de la República Argentina*. Santiago, 1890.

Lamas, Andrés. *Estudio histórico y científico del Banco de la Provincia de Buenos Aires*. Buenos Aires, 1886.

Larden, Walter. *Argentine Plains and Andine Glaciers*. London, 1911.

Latham, Wilfrid. *The States of the River Plate*. 2nd ed. London, 1868.

Latzina, Francisco. "El comercio argentino antaño y hogaño." In *Censo Agropecuario Nacional. La ganadería y la agricultura en 1908. Monografías*. Vol. III. Buenos Aires, 1909.

Le Long, J. "Les Pampas de la République Argentine," *Bulletin de la Société de Géographie*, 6th series, 15 (1878), 193–212.

Lemée, Carlos. *La agricultura y la ganadería en la República Argentina*. La Plata, 1894.

Link, Pablo. *Razas ovinas*. Buenos Aires, 1937.

Lix-Klett, Carlos. *Estudio sobre producción, comercio y finanzas de la República*. 2 vols. Buenos Aires, 1900.

Lord Brassey. *Voyages and Travels of Lord Brassey. From 1862 to 1894*. London, 1895.

MacCann, William. *Two Thousand Miles' Ride through the Argentine Provinces, Being an Account of the Natural Products of the Country and the Habits of the People*. 2 vols. London, 1853.

MacCorquodale, D. *The Argentine Revisited, 1881–1906*. Glasgow, 1909.

MacLeod, N. "The Life of a Sheep-farmer in the Argentine Republic," *Goodwords*, 12 (1871), 712–722.

Mannequin, M. T. "Les Provinces Argentines et Buenos Aires depuis leur Indépendence jusqu'à nos jours," *Journal des économistes*, August 1856.

Marmier, Xavier. *Buenos Aires y Montevideo en 1850*. Montevideo, 1967.

Martin de Moussy, J. A. Victor. *Description géographique et statistique de la Confédération Argentine*. Paris, 1860/73.

Maxwell, Daniel. *Planillas estadísticas de la exportación en los años desde 1849 a 1862 con algunas observaciones sobre ellas y la economía rural de nuestro país*. Buenos Aires, 1863.

Modrich, Guiseppe. *La Reppublica Argentina. Note di viaggio*. Milan, 1890.

Mulhall, M. G., and E. T. Mulhall. *Handbook of the River Plate Republics*. London and Buenos Aires, 1869, 1875, 1876, 1885, 1892.

Murray, Thomas. *The Story of the Irish in Argentina*. New York, 1919.

Napp, Ricardo. *La República Argentina*. Buenos Aires, 1876. English edn. *The Argentine Republic*. London, 1876.

Ogilvie, Campbell (ed.). *Argentine from a British Point of View and Notes of Argentine Life*. London, 1910.

Oldendorff, E. "Sheep Husbandry and Wool Production in the Argentine Republic," *Bulletin of the National Association of Wool Manufacturers*, 8 (1878), 224–230.

Olivera, Eduardo. *Miscelánea. Escritos económicos, administrativos, económico-rurales, agrícolas, ganaderos, exposiciones, discursos inaugurales y parlamentarios, viajes, correspondencia, historia y legislación*. 2 vols. Buenos Aires, 1910.

———. *Historia de la ganadería, agricultura e industrias afines en la República Argentina, 1515–1927*. Buenos Aires, 1928.

————. "Nuestra industria rural bajo su aspecto económico en 1867," *La Revista de Buenos Aires*, 15 (1868).

La Olivera, Soc. Anon. *Los establecimientos de la Sociedad Anónima La Olivera.* Buenos Aires, 1900.

Parish, Woodbine. *Buenos Aires y las provincias del Rio de la Plata.* Buenos Aires, 1958.

Patroni, Adrián. *Los trabajadores en la República Argentina.* Buenos Aires, 1897.

Peuchgaric, N. *La Plata de 1851 à 1854, relation des événements politiques, moeurs-coûtumes, caractère, éducation, gouvernement, commerce, etc.* Paris, 1857.

Pierrard, P. *Tableaux synoptiques du commerce des laines en Angleterre, 1888.* London, 1889.

Poole, B. *The Commerce of Liverpool.* London, 1854.

Quesada, Hector. *El crédito territorial en la República Argentina.* Buenos Aires, 1888.

Quesada, Sixto. *El Banco Hipotecario de la Provincia de Buenos Aires.* Buenos Aires, 1894.

Rancourt, E. de. *Fazendas et estancias: Notes de voyage sur la République Argentine.* Paris, 1901.

Rosas, Juan M. de. *Instrucciones a los mayordomos de estancias.* Buenos Aires, 1942.

Ross, Johnson. *Vacaciones de un inglés en la Argentina.* Buenos Aires, 1868.

Rumbold, Horace. *The Great Silver River. Notes of a Residence in Buenos Aires, 1880–1881.* London, 1887.

Ruiz, Dolores. *La Provincia de Buenos Aires o sean sus 81 partidos estudiada bajo la faz de la estadística, del comercio, agrícola, industrial, etc.* Buenos Aires, 1886.

Sauerbeck, A. *Production and Consumption of Wool.* London, 1878.

Seeber, Francisco. *Apuntes sobre la importancia económica y financiera de la República Argentina.* Buenos Aires, 1888.

Seymour, Richard. *Un poblador de las pampas. Vida de un estanciero de la frontera sudeste de Córdoba entre los años 1856 y 1868.* Buenos Aires, 1947.

Shaw, A. E. *Forty Years in the Argentine Republic.* London, 1907.

Tornquist, Ernesto. *The Economic Development of the Argentine Republic.* New York, 1918.

Uriarte, José (ed.). *Los Baskos en la Nación Argentina.* Buenos Aires, 1919.

Veber, A. "République Argentine," *La Revue Socialiste*, 13 (1891), 358–359.

Victory y Suárez, Bartolomé. *Cuestiones de interés público.* Buenos Aires, 1873.

Vicuña Mackenna, Benjamín. *La Argentina en el año 1855*. Buenos Aires, 1936.
Wiener, Charles. *Histoire du troupeau. Ses origines, ses croissements, son denombrement*. Paris, 1899.
Woodgate, C. F. *Sheep and Cattle Farming in Buenos Aires, with Sketch of the Financial and Commercial Position of the Argentine Republic*. London, 1876.
X. *Quinze jours au pays des cédules*. Gand, 1893.
Zeballos, Estanislao. *Descripción amena de la República Argentina*. 3 vols. Buenos Aires, 1881/1888.

Newspapers and Periodicals:

Anales de la Sociedad Rural Argentina. Vols. 1–14. Buenos Aires, 1867–1890.
American Economist. Vols. 7–19. New York, 1891–1897.
Annuaire des Deux Mondes. Vol. 7. Paris, 1856–1857.
Bankers' Magazine. Vols. 27–48. London, 1867–1888.
Bankers' Magazine. Vols. 12, 35. New York, 1857/1858, 1880/1881.
Board of Trade Journal. Vols. 3–24. London, 1887–1898.
Bradstreet's. Vols. 4–20. New York, 1881–1892.
The Brazil and River Plate Mail. 1863–1878. Continued as *The South American Journal*. 1879–1885. London, 1863–1885.
The British Packet and Argentine News. 1840–1848. Buenos Aires.
Bulletin of the National Association of Wool Manufacturers. Vols. 1, 2, 20, 28, 29. Boston, 1869, 1871, 1890, 1898, 1899.
Bulletin de la Société de Géographie Commerciale de Bordeaux. Vols. 3–14. Bordeaux, 1880–1890.
Le Courier de la Plata. Buenos Aires, 1867–1880.
The Economist. Vols. of 1843–1892. London.
L'Economiste Français. Vols. of 1873–1895. Paris.
Globus. Vols. 9–25. Hildburghaussen, 1866–1874.
Handels' Archiv. Vols. 55–63. Berlin, 1855–1863.
Hunt's Merchant Magazine. Continued as *The Merchant's Magazine and Commercial Review*. Vols. 6, 13, 21, 28, 40, 45. New York, 1842, 1845, 1849, 1853, 1859, 1861.
Revista Argentina. Vols. 1–13. Buenos Aires, 1868–1872.
Revue Britannique. Vols. of 1851 and 1866. Paris.
The Standard and River Plate News. Buenos Aires, 1870–1885.
La Tribuna. Buenos Aires, 1853–1861.
The Weekly Standard. Buenos Aires, 1870–1874.

Directories:

Almanaque agrícola, industrial y comercial de Buenos Aires. Years 1860, 1861, 1862, 1863.

Almanaque agrícola, pastoril e industrial de la República Argentina y de Buenos Aires. Buenos Aires, 1865.

Almanaque comercial y guía de forasteros para el Estado de Buenos Aires. Buenos Aires, 1855.

Almanaque-Guía de la Nación. Buenos Aires, 1871.

Almanaque Nacional para. . . . Years 1869 and 1871.

Almanaque para el año de Nuestro Señor 1861. Buenos Aires, 1861.

Almanaque Popular Doble, 1864. Buenos Aires, 1864.

Amorena, J. A. *Memorandum enciclopédico administrativo y comercial, descriptivo de Buenos Aires, capital de la República Argentina.* Buenos Aires, 1885.

El Avisador. Guía general de comercio y de forasteros. Years 1862–1867.

Gran Guía General de la República Argentina, 1873. Buenos Aires, 1873.

Gran Guía General Comercial de la República Argentina: estadística, agricultura, administración, 1878–1879. Buenos Aires, 1879.

Guía Kraft. Buenos Aires, 1890.

Guía mensual del comercio. Buenos Aires, 1882.

Kidd's guia mensual de los ferrocarriles, vapores, tramways y mensagerias. Buenos Aires, 1874.

Kunz y Cia. *Gran guía de la ciudad de Buenos Aires.* Buenos Aires, 1886.

Martinez y Cia. *Guida del Forastiere a Buenos Aires.* Buenos Aires, 1882.

Missolz, A. de. *Guía de comercio y de la industria de la ciudad y provincia de Buenos Aires.* Buenos Aires, 1881.

Pillado, Antonio. *Diccionario de Buenos Aires o sea Guía de Forasteros.* Buenos Aires, 1864.

Ruiz, Francisco. *Gran guía general del comercio de la República Argentina.* Years 1874, 1875, 1878/81.

Secondary Works

Abbot, G. *The Pastoral Age: A Re-examination.* Melbourne and Sidney, 1971.

Alvarez, Juan. *Estudio sobre las guerras civiles argentinas.* 3rd edn. Buenos Aires, 1938.

———. *Temas de historia económica argentina.* Buenos Aires, 1929.

Allende, Andrés. "Las fronteras del Estado de Buenos Aires," *Trabajos y comunicaciones,* vol. 1 (1949), 13–45.

———. "La ley de arrendamientos rurales del 21 de octubre de 1857 en la Provincia de Buenos Aires," *Trabajos y comunicaciones,* 18 (1968), 45–51.

———. *Historia del pueblo y partido de Lincoln en el siglo XIX: la conquista del oeste bonaerense.* La Plata, 1969.

Amaral, Samuel. "Comercio y crédito: el Banco de Buenos Aires (1822–
' 1826)," *América: Revista cuatrimestral de Asuntos Históricos,* 2, No. 4
(April, 1977), 9.
Archetti, Eduardo, and Kristi Stolen. *Explotación familiar y acumulación
de capital en el campo argentino.* Buenos Aires, 1975.
Astesano, Eduardo. *Rosas: Bases del nacionalismo popular.* Buenos Aires,
1960.
Bairoch, Paul. *Revolución industrial y subdesarrollo.* Mexico City, 1967.
Barba, Enrique. *Rastrilladas, huellas y caminos.* Buenos Aires, 1956.
Barba, E., et al. "La Campaña del Desierto y el problema de la tierra.
La ley de 1878 y su aplicación en la Provincia de Buenos Aires." In
*Segundo Congreso de Historia de los pueblos de la Provincia de Buenos
Aires.* La Plata, 1974.
Barnard, Alan. *The Australian Wool Market, 1840–1900.* Melbourne, 1958.
Barran, José, and Benjamin Nahum. *Historia rural del Uruguay moderno.*
Montevideo, 1967–1972.
Bauer, Arnold. "Rural Workers in Spanish America: Problems of Peonage
and Oppression," *Hispanic American Historical Review,* 59 (1979), 34–
63.
Bejarano, Manuel. "Inmigración y estructuras tradicionales en Buenos
Aires (1854–1930)." In T. Di Tella and T. Halperin Donghi (eds.). *Los
fragmentos del poder.* Buenos Aires, 1968.
Beyhaut, Gustavo, et al. "Los inmigrantes en el sistema ocupacional
argentino." In T. Di Tella et al., *Argentina, sociedad de masas.* Buenos
Aires, 1965.
Bidabehere, Fernando. *Bolsas y mercados de comercio en la República Ar-
gentina.* Buenos Aires, 1930.
Boglich, José. *La cuestión agraria en la Argentina.* Buenos Aires, 1964.
Bosco, Eduardo. *El gaucho a través de los testimonios extranjeros, 1773–1870.*
Buenos Aires, 1947.
Brown, Jonathan. *A Socioeconomic History of Argentina, 1776–1860.* Cam-
bridge, 1979.
Bugueño, José. *Contribución al estudio de la fundación y desarrollo del pueblo
de San Antonio de Areco.* S. A. de Areco, 1936.
Butlin, N. G. *Investment in Australian Economic Development, 1861–1900.*
Cambridge, 1964.
Cárcano, Miguel A. *Evolución histórica del régimen de la tierra pública 1810–
1916.* Buenos Aires, 1917.
Carreño, Virginia. *Estancias y estancieros.* Buenos Aires, 1968.
Carretero, Andres. "Contribución al conocimiento de la propiedad rural
en la Provincia de Buenos Aires para 1830," *Boletín del Instituto de
Historia Argentina Dr. E. Ravignani,* 2nd. series, 13 (1970).

————. *La propiedad de la tierra en la época de Rosas.* Buenos Aires, 1972.

Casa Pardo. *Nuestras estancias.* Buenos Aires, 1968.

Casarino, Nicolás. *El Banco de la Provincia de Buenos Aires en su primer centenario, 1822–1922.* Buenos Aires, 1922.

Chiaramonte, José C. *Nacionalismo y liberalismo económicos en Argentina (1860–1880.)* Buenos Aires, 1971.

Coghlan, Eduardo. *Los irlandeses.* Buenos Aires, 1970.

Coni, Emilio. *El gaucho. Argentina. Brasil. Uruguay.* Buenos Aires, 1969.

Cortés Conde, Roberto. *El progreso argentino, 1880–1914.* Buenos Aires, 1979.

Cortés Conde, Roberto, Tulio Halperin Donghi, and Haydée Gorostegui de Torres. *Evolución del comercio exterior argentino. 1a. parte, 1864–1930. Exportaciones.* Mimeo. Buenos Aires, 1965.

Cuccorese, Horacio. *Historia del Banco de la Provincia de Buenos Aires.* Buenos Aires, 1972.

————. "Historia sobre los orígenes de la Sociedad Rural Argentina," *Humanidades,* 35 (1960), 23–53.

Díaz, Benito. *Juzgados de paz de campaña de la Provincia de Buenos Aires (1821–1854).* La Plata, 1959.

Duncan, R., and I. Rutledge. *Essays on the Development of Agrarian Capitalism in the Nineteenth and Twentieth Centuries.* New York, 1977.

Ferns, Henry S. *Britain and Argentina in the Nineteenth Century.* Oxford, 1960.

Ferrer, Aldo. *La economía argentina.* Buenos Aires, 1963.

Flichman, Guillermo. *La renta del suelo y el desarrollo agrario argentino.* Mexico City, 1977.

Florescano, Enrique (ed.). *Haciendas, latifundios y plantaciones en América Latina.* Mexico, 1975.

Fohlen, Claude. *L'industrie textile au temps du Second Empire.* Paris, 1956.

Ford, A. G. *The Gold Standard, 1880–1914: Britain and Argentina.* Oxford, 1962.

Gallo, E., and R. Cortés Conde. *La república conservadora.* Buenos Aires, 1972.

Geller, Lucio. "El crecimiento industrial argentino hasta 1914 y la teoría del bien primario exportable." In M. Giménez Zapiola, *El régimen oligárquico.* Buenos Aires, 1975.

Giberti, Horacio. *El desarrollo agrario argentino: estudio de la región pampeana.* Buenos Aires, 1964.

————. *Historia económica de la ganadería argentina.* Buenos Aires, 1961.

Gori, Gastón. *La pampa sin gaucho. Influencia del inmigrante en la transformación de los usos y costumbres en el campo argentino en el siglo XIX.* Buenos Aires, 1952.

———. *Vagos y malentretenidos.* Santa Fe, 1951.

Gorostegui de Torres, Haydée. *Argentina. La organización nacional.* Buenos Aires, 1972.

Gutelman, Michel. *Structures et réformes agraires: Instruments pour l'analyse.* Paris, 1974.

Halperin Donghi, Tulio. "La expansión ganadera en la campaña de Buenos Aires." In T. Di Tella and Tulio Halperin Donghi. *Los fragmentos del poder.* Buenos Aires, 1968.

———. "La expansión de la frontera de Buenos Aires (1810–1852)." In Alvaro Jara (ed.). *Tierras Nuevas.* Mexico City, 1969.

———. *José Hernández y sus mundos.* Buenos Aires, 1985.

———. *Revolución y guerra. Formación de una élite dirigente en la Argentina criolla.* Buenos Aires, 1972.

Hernando, Diana. *Casa y familia: Spatial Biographies in Nineteenth Century Buenos Aires.* Ph.D. diss., University of California, Los Angeles, 1973.

Iriart, Michel. *Corsarios y colonizadores vascos.* Buenos Aires, 1945.

Jones, Charles. *British Financial Institutions in Argentina, 1860–1914.* Ph.D. diss., Cambridge University, 1973.

Jones, Wilbur. "The Argentine British Colony in the Time of Rosas," *Hispanic American Historical Review,* 40, No. 1 (February, 1960), 90–97.

Korner, Karl. "El Cónsul Zimmerman, su actuación en Buenos Aires, 1815–1847," *Boletín del Instituto de Historia Argentina Dr. E. Ravignani,* 2nd. series, No. 11–13 (1966), 3–166.

Korol, J. C., and H. Sabato. *Como fue la inmigración irlandesa a la Argentina.* Buenos Aires, 1981.

Kroeber, Clifton. *The Growth of the Shipping Industry in the Rio de la Plata Region, 1794–1860.* Madison, 1957.

Laclau, Ernesto. "Modos de producción, sistemas económicos y población excedente: Aproximación histórica a los casos argentino y chileno," *Revista Latinoamericana de Sociología,* 5 (1969), 276ff.

Lebedinsky, Mauricio. *Estructura de la ganadería.* Buenos Aires, 1967.

Lebrun, Pierre. *L'industrie de la laine à Verviers pendant le XVIIIe et le début du XIXe siècle.* Paris, 1948.

Levene, Ricardo. *Historia de la Provincia de Buenos Aires y formación de sus pueblos.* La Plata, 1941.

Lewis, Colin. "Problems of Railway Development in Argentina, 1857–1890," *Interamerican Economic Affairs,* 22, No. 2 (August, 1968), 55–75.

Lynch, John. *Argentine Dictator, Juan Manuel de Rosas, 1829–1852.* Oxford, 1981.

Macchi, Manuel. *El ovino en la Argentina*. Buenos Aires, 1974.

Mendoza, Prudencio. *Historia de la ganadería argentina*. Buenos Aires, 1928.

Minola, Jose. *Historia del lanar*. Buenos Aires, 1976.

Moncaut, Carlos. *Biografía del río Salado de la Provincia de Buenos Aires*. La Plata, 1966.

———. *Estancias bonaerenses*. City Bell, 1977.

———. *Pampas y estancias*. City Bell, 1978.

Montoya, Alfredo. *La ganadería y la industria del salazón de carnes en el período 1810–1862*. Buenos Aires, 1971.

———. *Historia de los saladeros argentinos*. Buenos Aires, 1956.

Morrissey, Sylvia. "The Pastoral Economy, 1821–1850." In James Griffin (ed.). *Essays in Economic History of Australia*. 2nd ed. Queensland, 1970.

Newton, Jorge. *Cabañas argentinas: la historia de la ganadería nacional con la introducción de las primeras tropas del Brasil hasta la actualidad*. Buenos Aires, 1970.

———. *Diccionario biográfico del campo argentino*. Buenos Aires, 1972.

Newton, Jorge, and Lily Sosa de Newton. *Historia de la Sociedad Rural Argentina*. Buenos Aires, 1966.

———. *Historia del Jockey Club de Buenos Aires*. Buenos Aires, 1966.

Nichols, M. W. *The Gaucho. Cattle Hunter. Cavalryman. Ideal of Romance*. Durham, 1943.

Oddone, Jacinto. *La burguesía terrateniente argentina*. Buenos Aires, 1967.

Olarra Jimenez, V. *Evolución monetaria argentina*. Buenos Aires, 1967.

Ortiz, Ricardo. *Historia económica de la Argentina*. 2 vols. Buenos Aires, 1964.

Panettieri, Jose. *La crisis ganadera: Ideas en torno a un cambio en la estructura económica y social del país (1866–1871)*. La Plata, 1963.

———. *Los trabajadores en tiempos de la inmigración masiva en Argentina (1870–1910)*. La Plata, 1965.

Piñero, Norberto. *La moneda, el crédito y los bancos en la Argentina*. Buenos Aires, 1921.

Poulain, Gaston. *La délainage et sa capitale Mazamet*. Mazamet, 1951.

Les Problèmes agraires des Amériques Latines. Colloques Internationaux du CNRS, Paris, 11–16 October 1965. Paris, 1967.

Pucciarelli, Alfredo. *El capitalismo agrario pampeano, 1890–1930*. Buenos Aires, 1986.

Quintero Ramos, A. S. *A History of Money and Banking in Argentina*. Rio Piedras, Puerto Rico, 1965.

Reber, Vera. *British Mercantile Houses in Buenos Aires, 1810–1880*. Cambridge, Mass., 1979. (Citations in the text are from the original Ph.D. dissertation, University of Wisconsin, 1972.)

Roberts, S. H. *The Squatting Age in Australia, 1835–1847*. Melbourne, 1964.

Rodríguez Molas, Ricardo. *Historia social del gaucho*. Buenos Aires, 1968.

Rögind, William. *Historia del Ferrocarril del Sud, 1861–1936*. Buenos Aires, 1937.

Sabato, Hilda. "La formación del mercado de trabajo en Buenos Aires, 1850–1880," *Desarrollo Económico*, 24, No. 96 (January–March, 1985), 561–592.

Sábato, Jorge F. *La clase dominante en la Argentina moderna: Formación y características*. Buenos Aires, 1988.

Sáenz Quesada, María. *Los estancieros*. Buenos Aires, 1980.

Sánchez Albornoz, Nicolás. "Rural Population and Depopulation in the Province of Buenos Aires: 1869–1960." In International Economic History Conference, IV, Bloomington, Indiana, 1968: *Proceedings, section 5. Historical Demography*. Winnipeg, 1970.

Sbarra, Noel. *Historia del alambrado en la Argentina*. 2nd ed. Buenos Aires, 1964.

Scardin, F. *La estancia argentina*. Buenos Aires, 1908.

Scobie, James. *La lucha por la consolidación de la nacionalidad argentina, 1852–1862*. Buenos Aires, 1964.

———. *Revolución en las pampas: Historia social del trigo argentino, 1860–1910*. Buenos Aires, 1968.

Slatta, Richard. *Gauchos and the Vanishing Frontier*. Lincoln, Neb., 1983.

Shaw, Edward. *An Economic History of Australia*. Cambridge, 1930.

Shannon, Fred. *The Farmer's Last Frontier: Agriculture, 1860–1897*. New York, 1945.

Sireau, Albert. *Culture et peuplement. Evolution du Municipio de Mercedes dans la Province de Buenos Aires*. 1965.

Sociedad Rural Argentina. *Tiempos de epopeya, 1866–1966*. Buenos Aires, 1966.

Svec, William. *A Study of the Socioeconomic Development of the Modern Argentine Estancia, 1852–1914*. Ph.D. diss., University of Texas at Austin, 1966.

Toulemonde, Jacques. *Naissance d'une métropole: Histoire économique et sociale de Roubaix et Tourcoing au XIXe siècle*. Tourcoing, 1966.

Valencia de Placente, M. E. *La política de tierras públicas después de Caseros 1852–1871—Pcia. de Buenos Aires*. Ph.D. diss., University of La Plata, 1983.

Vázquez Presedo, V. *Estadísticas históricas argentinas comparadas*. Buenos Aires, 1971.

Villafañe Casal, M. T. *La mujer en la pampa, siglos XVIII y XIX*. La Plata, 1958.

Zalduendo, Horacio. *Libras y rieles*. Buenos Aires, 1975.

Index